Organic Electronics 2

Series Editor
Robert Baptist

Organic Electronics 2

Applications and Marketing

Thien-Phap Nguyen

iSTE

WILEY

First published 2021 in Great Britain and the United States by ISTE Ltd and John Wiley & Sons, Inc.

ISTE Ltd
27-37 St George's Road
London SW19 4EU
UK

www.iste.co.uk

John Wiley & Sons, Inc.
111 River Street
Hoboken, NJ 07030
USA

www.wiley.com

Library of Congress Control Number: 2021939365

British Library Cataloguing-in-Publication Data
A CIP record for this book is available from the British Library
ISBN 978-1-78630-610-4

Contents

Introduction

Organic electronics can be defined as a branch of general electronics that focuses on studying the properties and applications of organic semiconductors. These materials, referred to in common usage as plastics, are, in fact, a distinct class separate from that of ordinary plastic materials. They may be small molecules, or conjugated polymers, which have an electronic structure comparable to that of conventional semiconductors. Therefore, their physical properties are very similar to the properties of these latter materials. However, organic materials are generally amorphous and their electrical conductivity is much lower than that of traditional semiconductors. As a result, they have been neglected for a long time despite the discovery of some of their notable physical properties, such as the discovery of electroluminescence in anthracene in the 1960s.

In all scientific or technological disciplines, an idea, concept or creation can arise with a change or progression and profoundly alter the course of the evolution of the discipline – and sometimes even the course of the history of science. Such was the case in 1947 when the researchers John Bardeen, William Shockley and Walter Brattain invented the transistor, a device that would revolutionize traditional electronics. This invention is what first allowed for the design of circuits that provided logical functions, then for these circuits to be integrated into complex systems capable of handling sophisticated tasks, and finally for these miniaturized systems to be incorporated into the portable electronic devices that we now use every day. A similar progression occurred in organic electronics in 1987, when Tang and Van Slyke used the evaporation of small molecules in a vacuum to create diodes that emitted light in the form of a thin film that operated with a turn-on voltage lower than 10 V. Their work demonstrated that, in practice, organic devices are suitable for optoelectronic applications in the same way as their inorganic counterparts. A few years later, Burroughs *et al.* (1994) demonstrated that by using conjugated polymers deposited as thin films from a solution, they could also produce light-emitting diodes with a low operating voltage. This work paved the way for film depositing techniques using solutions and led to the production of devices through printing developed later in laboratories. For this domain, we will use the descriptors "flexible" or "printed" to indicate the discipline of organic electronics. Advances in research have been very rapid and have shown remarkable results, not only in the understanding of physical processes in materials and components but also for the creation of new electronic devices now available on the market. Today, organic electronics has become its own field, one that will prove to be important technologically and economically in the near future.

In France, education on organic electronics at the university level is developing, but remains limited in comparison with other European countries. It should be noted that these courses are generally offered in universities that carry out industrial research and development activities on organic electronic materials and devices. With regard to the books written on this discipline, there are very few titles available in French.

This book, divided into two volumes, was written with the goal of introducing organic electronics to students and researchers who are interested in this new discipline. It is organized into the following sections: Chapter 1 of Volume 1 provides an overview of the basic notions of the theory of traditional semiconductors, of which some of the material presented will be used later on. The materials used for the creation of devices are described in Chapter 2. The physical processes, which take place in the volume and interface of the layers of devices, are presented and explained in Chapters 3, 4 and 5. Volume 2 presents the primary applications of organic materials in optoelectronic devices. The first chapter of this volume centers on organic light-emitting diodes (OLEDs), the second on organic solar cells (OSCs or OPVs) and the third on organic field-effect transistors (OFETs). Chapter 4 deals with the practical and economic aspects of the industrialization of organic components. It also includes a discussion on the environmental aspects of the use of organic materials and devices.

As the title indicates, this book is not intended to provide a detailed and complete description of the materials, physical processes and applications involved in organic electronics. This would be far beyond the scope of what can be addressed within a single book. The choice of topics and of the extent to which the subjects are discussed were made by the author, based on his own experience from research and as an instructor. Interested readers will be able to find more detailed discussions on certain topics using the references provided. A list of acronyms used in the text is also included at the end of this book.

I would like to thank the people who devoted their time to proofreading and providing comments and suggestions, which allowed me to improve the writing of this book and the way in which certain areas are presented: Philippe Lerendu, Serge Lefrant, René Leparoux, Agnès Bournigal-Giret, Maxime Bayle and Jean-Luc Duvail. I would also like to thank ISTE for proposing this project, which I hope will help to arouse public interest in the new and promising possibilities offered by organic electronics in the scientific and technological communities in France and around the world.

1

Organic Light-Emitting Diodes

1.1. Introduction

According to the International Energy Agency (IEA), worldwide electricity consumption in 2015 represented 18% of total energy consumption, just behind oil at 41%. Almost one-fifth of electricity consumption is devoted to lighting, whether public, private, domestic or industrial. Until recently, incandescent lamps or fluorescent tubes were universal sources of lighting, with low efficiency and short lifetimes. These lamps have gradually been replaced by new inventions in the field of lighting, such as compact fluorescent lamps, *semiconductor light-emitting diodes* (LEDs), and then *organic light-emitting diodes* (OLEDs). These sources not only have good optical performance, but are also economical to use, as their electricity consumption is low compared to older generation lamps under the same lighting conditions.

OLEDs result from organic electronic technology and symbolize a new concept of lighting. This concept is based on a modern design, presenting evolving shape and volume, and on energy saving and sustainable development. Indeed, OLED lamps are composed of emitting surfaces, not point sources such as traditional lamps. In addition, they are lightweight and produced on thin substrates, enabling them to be mounted at the desired locations without any particular constraints. The materials used to manufacture the devices are available, non-toxic and not harmful to nature.

In addition to their use in lighting, OLEDs are also used in display and visualization devices. Their characteristics make it possible to design screens of higher quality than that of conventional technologies, with advantages due to the nature of the organic materials used: they are lightweight and have large active surfaces and low manufacturing costs. In this chapter, we will study the characteristics of OLEDs and their operations in comparison with conventional semiconductor (SC) LEDs.

1.2. Reminders on optics

1.2.1. *Photometry and radiometry*

When we use a lamp to light a room, we can distinguish between the values associated with the light source (the lamp) and the receiver (the room). These values are expressed in two different ways, depending on the perspective chosen. From a physical perspective, we consider light as a ray of wavelength λ, and the values associated with this ray are called *radiometric quantities*. From a vision perspective, light is associated with the perception and sensitivity of the human eye. Each ray of wavelength λ is perceived by the eye as a visible color. The associated values are called *photometric quantities*. In the definitions of the values, we will specify the units used in radiometry and photometry.

Let us consider a point source emitting a light beam of flux $\Delta\Phi$ and solid angle $\Delta\Omega$. The source intensity in the flux direction is defined by:

$$\Delta I = \frac{\Delta\Phi}{\Delta\Omega} \qquad [1.1]$$

The unit of *energy flux* is the watt (W), and that of *luminous flux* is the lumen (lm). The unit of radiant intensity is the watt per steradian (W/sr) and that of luminous intensity is the candela (cd).

The *luminance* of a light source is defined as the source intensity per unit area in a given direction (Δ). It characterizes the source seen by an observer whose eye is placed in direction (Δ). We obtain:

$$L = \frac{\Delta I}{\Delta S \times \cos\alpha} \qquad [1.2]$$

where ΔI is the intensity of the source, $\Delta S \times \cos\alpha$ represents the apparent surface of the source seen by the observer and α is the angle formed by the normal to the surface of the source and direction (Δ). The photometric unit of luminance is the *candela per square meter (cd/m²)*. The radiometric unit of luminance (also called *radiance*) is the *watt per square meter per steradian (W/m²/sr)*.

For the receiver, we define the *illuminance* or the *irradiance* as the incident light flux per unit area. Its unit is the watt per square meter (W/m²). The corresponding photometric unit is the lumen per square meter (lm/m²) or the lux (lx).

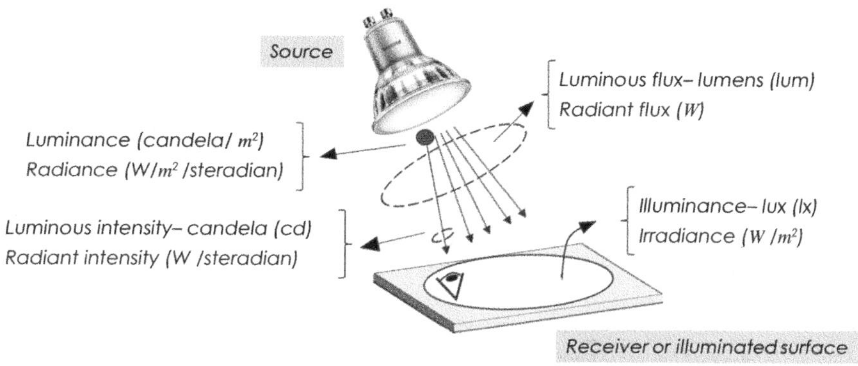

Figure 1.1. *Parameters of light sources and receivers*

Perception of light varies from one individual to another. For the same average observer, perception is maximum for the wavelength $\lambda = 555$ nm and the perceived sensation of luminosity is different when the light spectrum varies. To compare the perception of different light sources, we define the *luminous efficacy* or the *visibility factor* as the ratio of the visible light flux to the power absorbed by the source. It is expressed in lumen per watt (lm/W).

1.2.2. *Colors*

1.2.2.1. *Chromaticity diagram*

The visible light spectrum approximately covers the range of wavelengths from 400 to 800 nm. There are three *primary colors*, whose wavelengths are, respectively, $\lambda(R) = 700$ nm for red, $\lambda(V) = 546$ nm for green and $\lambda(B) = 435$ nm for blue. Any radiation of the visible spectrum corresponds to a color that can be obtained from a mixture of the three primary colors. For this presentation, we use components that are said to be *trichromatic*, defined by the following relations:

$$r = \frac{R}{R+V+B}, v = \frac{V}{R+V+B}, b = \frac{B}{R+V+B} \quad [1.3]$$

We then deduce:

$$r + v + b = 1 \quad [1.4]$$

However, using these components presents some mathematical disadvantages, so new, fictitious primary colors, X, Y, Z, are proposed to replace them. These colors are mathematical transforms of RGB (red, green and blue) radiations. They take account of the notion of luminance, which introduces the notion of human vision, and the associated new components are defined by:

$$x = \frac{X}{X+Y+Z}, y = \frac{Y}{X+Y+Z}, z = \frac{Z}{X+Y+Z} \quad [1.5]$$

We obtain:

$$x + y + z = 1 \quad [1.6]$$

This is the equation of a plane (P) in space. An arbitrary color, C, is represented in the plane by a point, called the *color point*, of coordinates (x, y, z). Yet only two of the three coordinates, x and y, for example, need to be known in order to determine the color point (as $z = 1 - x - y$). The diagram is presented in two dimensions with the x coordinate on the abscissa and y on the ordinate. This diagram is the *chromaticity diagram of the CIE* (*Commission Internationale de l'Eclairage* – International Commission on Illumination). The reference color point corresponds to the color white. Its coordinates in the (x, y) plane are $x = y = \frac{1}{3}$. It is located at the center of the diagram. The points of arbitrary color are located between the center and the edges. The dominant color is red when $x > y$, green when $y > x$ and blue when $z > x, y$.

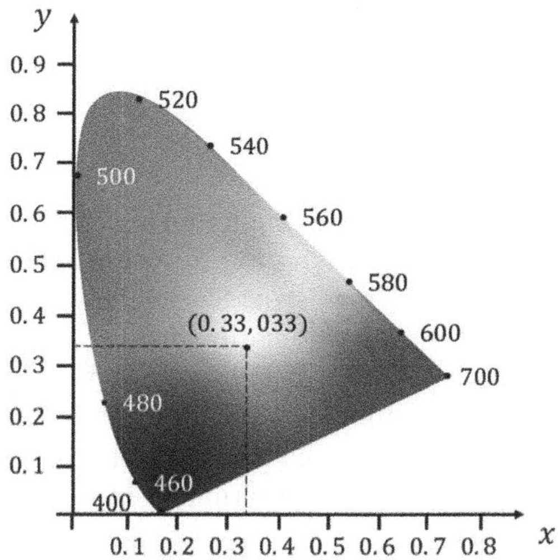

Figure 1.2. *CIE chromaticity diagram. For a color version of this figure, see www.iste.co.uk/nguyen/electronics2.zip*

1.2.2.2. *Color temperature*

To characterize the color of an incandescent source, we use the *color temperature* (*T*) parameter, defined as the temperature of a black body whose radiation color is identical or close to that of the source. In other words, it is the temperature at which a black body needs to be (thermally) heated to obtain the same color as the emitter. The temperature of solar radiations is approximately 5,800 K when the sun is at its zenith. Those of daylight vary between 4,000 and 7,500 K. In lighting, white light is favored for visual comfort. Two shades of white can be distinguished: *warm white*, with a low color temperature (< 3,500 K), yellowish in tone; and *cool white*, with a high color temperature (> 4,500 K), bluish in tone.

1.2.2.3. *Color rendering index*

To evaluate the colorimetric quality of a light source, we use a parameter called the *color rendering index* (CRI), a number with a value that varies between 0 and 100 and which is used to evaluate the source's ability to reproduce the colors of an object illuminated by this source. The index allocated to a source is compared to that of a reference source with the same color temperature (generally the black body) that has an index of 100. For a high-quality white-light source, the CRI is greater than 85.

1.3. OLED operating principle

1.3.1. *P–N junction LED*

1.3.1.1. *Operating principle*

When we apply a voltage, V_0, to the terminals of a forward-biased P–N junction, the electrons of the N-type SC diffuse across the space charge region (SCR) in the vicinity of the P-type SC, and the holes of the P-type SC diffuse into the N-type SC. These minority charge carriers recombine with the majority carriers of the SC in which they are to be found after diffusion. The recombination region is limited by the diffusion length of the holes, L_p, in the N-type SC and the diffusion length of the electrons, L_n, in the P-type SC. The active emission region is determined by these lengths from the limits of the junction SCR.

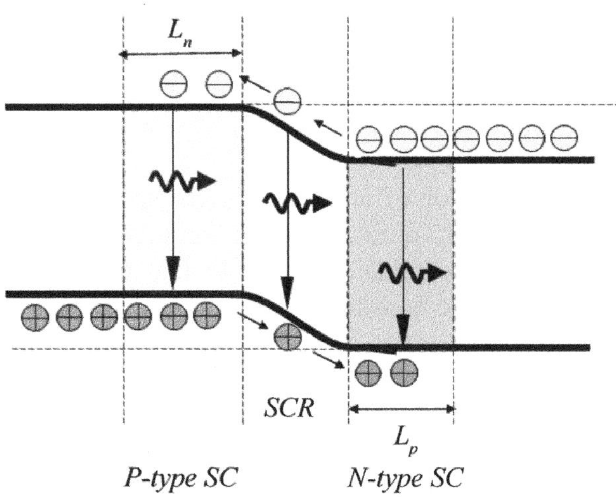

Figure 1.3. *P–N junction LED operating principle*

1.3.1.1.1. Recombination of excess carriers

In an ideal junction, each electron injected from the N-type SC into the P-type SC generates a photon. If the recombination is radiative, the energy of the photon emitted is equal to the energy of the SC gap:

$$E_G = h\,\nu \quad [1.7]$$

Let n_0 and p_0 be the concentrations of electrons and holes of the SC at equilibrium. When the SC is in non-equilibrium through an internal process, the carrier concentrations become $n = n_0 + dn$ and $p = p_0 + dp$. The *spontaneous recombination rate* is defined as the speed at which the concentration of carriers decreases. When an exciton disappears, an electron and a hole disappear. We can therefore write:

$$R = -\frac{dn}{dt} = -\frac{dp}{dt} \qquad [1.8]$$

The recombination probability of a CB electron with a VB hole is proportional to the concentration of VB holes. We can therefore write $R \propto p$. The same reasoning for the recombination of a VB hole with a CB electron leads to $R \propto n$. Finally, we can write:

$$R = B\, n\, p \qquad [1.9]$$

where B is a proportionality coefficient called the *bimolecular recombination coefficient*. For use in light emission applications, SCs need to have a high recombination coefficient (10^{-11}-10^{-9} cm^3s^{-1}).

1.3.1.1.2. Emission of photons

When a voltage, V_0, is applied under forward bias at the junction, the potential barrier between two SCs decreases by a value equal to $|e|V_0$. The diffusion current of majority carriers from one SC to the other proportionally increases to $\exp\left(\frac{|e|V_0}{kT}\right)$. The diffused carriers originating from an SC are minority carriers and may recombine with the majority carriers of the other SC. If the recombination is radiative, there will be an emission of photons by the junction.

Let n_0 and p_0 be the concentrations of electrons and holes of the SC at thermal equilibrium, n and p the concentrations of electrons and holes under electrical excitation and Δn and Δp the concentrations of excess electrons and holes. The recombination rates of excess carriers are written as:

$$\begin{aligned} R' &= B\, n\, p = B\,(n_0 + \Delta n)\,(p_0 + \Delta p) \\ &= B\, n_0\, p_0 + B\,(n_0\, \Delta p + p_0\, \Delta n + \Delta p \Delta n) \end{aligned}$$

By overlooking the second-order term, we obtain the number of recombinations due to the excess carriers created by electrical excitation:

$$\Delta R = R' - B\, n_0\, p_0 \sim B(n_0\, \Delta p + \, p_0\, \Delta n) \qquad [1.10]$$

We can write this term in the form:

$$\Delta R \sim \frac{\Delta p}{\tau_{rp}} + \frac{\Delta n}{\tau_{rn}} \qquad [1.11]$$

In this expression, we have:

$$\tau_{rp} \ = \frac{1}{B\, n_0} \qquad [1.12a]$$

$$\tau_{rn} \ = \frac{1}{B\, p_0} \qquad [1.12b]$$

where τ_{rp} is the radiative lifetime of the holes and τ_{rn} is the radiative lifetime of the electrons. Other physical processes that do not produce photon emission are referred to as non-radiative recombination and contribute to the overall lifetime, τ, of charge carriers.

We can write:

$$\frac{1}{\tau} = \frac{1}{\tau_r} + \frac{1}{\tau_{nr}} \qquad [1.13]$$

where τ_r and τ_{nr} are the radiative and non-radiative lifetimes, respectively, of the carriers (holes or electrons).

The carrier lifetimes are used to evaluate the efficiency of the emission of photons by the SC under electrical excitation. We can define a parameter called the *internal quantum efficiency* using the ratio:

$$\eta_r = \frac{1/\tau_r}{1/\tau} = \frac{\tau_{nr}}{\tau_r + \tau_{nr}} \qquad [1.14]$$

It represents the ratio of radiative recombination to the overall recombination (radiative and non-radiative).

1.3.1.2. *Materials and colors*

The SCs and their components for the LED emitter are chosen according to the material emission color, that is, according to their gap. The most commonly used materials are compounds of elements belonging to column III (boron, aluminum, gallium and indium) and to column V (nitrogen, phosphorus, arsenic and antimony). By varying the composition of the compounds, we can adjust their gap and thus modify the color emitted. The LEDs marketed can emit wavelengths from 1.43 eV (infrared for the compound GaAs) up to 5.90 eV (ultraviolet for the compound AlN). Diodes emitting the color blue are difficult to produce because their efficiency is low and the compounds are generally unstable. A Japanese team led by three researchers, Akasaki, Amano and Nakamura (Nakamura 1998), succeeded in synthesizing a compound of indium gallium nitride (InGaN) that emitted a stable blue light of high efficacy (300 lm/W). This discovery was awarded the Nobel Prize in Physics in 2014.

It is not at present possible, however, to find compounds emitting white light. Nevertheless, where lighting is concerned, white light provides essential visual comfort as it is close to natural light. To obtain a spectrum covering the entire visible range with a relatively constant intensity, three techniques are available:

– a combination of the three primary colors, red, green and blue, which produces white light. The main difficulty with this technique is the dosage of the intensity of emission colors required to obtain a homogeneous white without any dominant tint or shade (see Figure 1.4a);

– use of red, green and blue phosphors excited by UV radiations from an emitter. The resulting emission spectrum practically covers the visible spectrum, enabling white light to be obtained. The color emission efficiency is low owing to the chromophore excitation process, which constitutes an energy transfer (see Figure 1.4b);

– use of a blue emitter associated with a yellow-emitting phosphor. This technique is the most widely used in industry, as it uses only two materials to obtain a white emission (see Figure 1.4c).

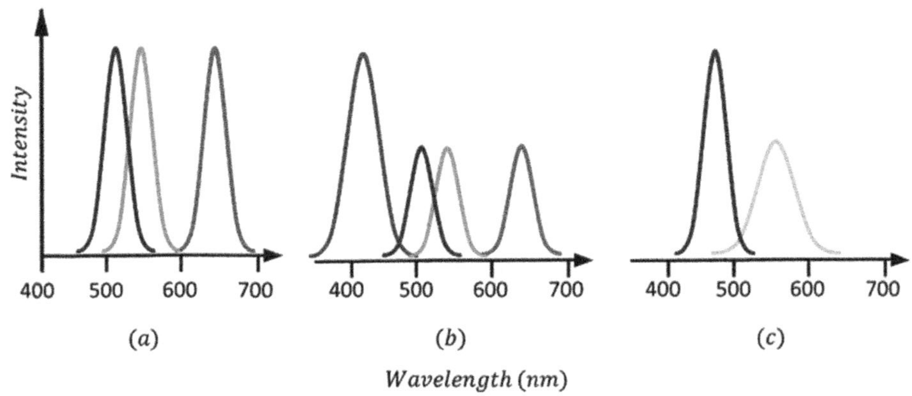

Figure 1.4. *Principle of obtaining white light from chromophores or emitting materials. For a color version of this figure, see www.iste.co.uk/nguyen/electronics2.zip*

1.3.1.3. *Extraction of light emitted*

The architecture of an LED must, in principle, be designed such that it enables all of the photons created in the diode to be extracted towards the external medium. In reality, because of the optical processes at the SC/external medium interface, few photons are collected on the outside of the diode.

1.3.1.3.1. Power losses at the semiconductor/air interface

Let us consider an SC of refractive index n, which emits photons to ambient air of index $n_0 = 1$. First, for rays perpendicular to the SC surface, the reflection coefficient is:

$$R = \frac{(n-1)^2}{(n+1)^2} \qquad [1.15]$$

For an SC of refractive index $n = 3$, the value of R is 0.25. The percentage of photons reflected on the surface of the SC is 25%.

For rays of incidence of angle θ_i with respect to the normal to the surface, the angle of refraction, θ_r, in the external environment is given by the Snell–Descartes law:

$$n_0 \sin\theta_r = n \sin\theta_i \qquad [1.16]$$

As $n > n_0$, the incident ray will be entirely reflected on the surface of the SC when $\theta_r \geq \pi/2$. The limit incident angle, θ_l, is such that:

$$n_0 \, sin \left(\frac{\pi}{2}\right) = n \, sin \, \theta_l \qquad [1.17]$$

and:

$$\theta_l = arc \, sin \left(\frac{n_0}{n}\right) = arc \, sin \left(\frac{1}{n}\right) \qquad [1.18]$$

If the incident angle is greater than θ_l, the ray will be reflected on the surface of the SC and will not be released from the sample. It can then be absorbed by the material or be released, attenuated, to the external environment after several reflections.

The rays emitted in the SC and contained in the volume of a cone, called an *extraction cone*, with a half-angle at the top, θ_l, will be released into the medium external to the SC.

Let Ω_l be the solid angle corresponding to the extraction cone. We obtain:

$$\Omega_l = 2\,\pi \int_0^{\theta_l} sin \, \theta \, d\theta = 2\,\pi \,(1 - cos \, \theta_l) \qquad [1.19]$$

For a small θ_l angle, we can write:

$$\Omega_l \sim 2\,\pi \left(1 - 1 + \frac{\theta_l^2}{2}\right) = 2\,\pi \frac{1}{2\,n^2} = \frac{\pi}{n^2} \qquad [1.20]$$

The rays emitted by the SC cover a solid angle equal to $\Omega = 4\pi$ (steradians). The ratio of the solid angle of the rays emitted inside the SC to that collected in ambient air is:

$$\frac{\Omega_l}{\Omega} \sim \frac{\pi}{4\,\pi\, n^2} = \frac{1}{4\, n^2} \qquad [1.21]$$

Let P_s be the optical power of the light source. The power available in the extraction cone of this source is:

$$P_d = \frac{\Omega_l}{\Omega} \, P_s = \frac{P_s}{4\, n^2} \qquad [1.22]$$

For an SC of refractive index $n = 3$, the power available in the extraction cone is approximately 2.70% of the power of the rays emitted in the SC.

However, the power actually collected on the outside of the diode needs to take account of the reflection on the surface of the SC, whose coefficient, *R*, is given by expression [1.16]. The transmission of the rays emitted is determined by the transmittance, *T*:

$$T = 1 - R = 1 - \frac{(n-1)^2}{(n+1)^2} = \frac{4\,n}{(n+1)^2} \qquad [1.23]$$

The power actually available on the outside is:

$$P_e = \frac{P_S}{4\,n^2} \times \frac{4\,n}{(n+1)^2} = \frac{1}{n\,(n+1)^2}\,P_S \qquad [1.24]$$

The *optical efficiency* is defined by the expression:

$$\eta_O = \frac{P_e}{P_S} = \frac{1}{n\,(n+1)^2} \qquad [1.25]$$

For $n = 3$, the optical efficiency of the diode will be $\eta_O \sim 2\%$.

1.3.1.3.2. Improving extraction efficiency

According to expression [1.25], the optical efficiency depends on the refractive index, *n*, of the SC and we can improve it by changing the material and choosing an appropriate one. This is not a solution in practice. We can, however, change the external surrounding medium of the SC to modify the limit incident angle, θ_l, because the latter depends on the refractive index, n_0, of the medium external to the SC.

According to expression [1.18], we obtain:

$$\theta_l = arc\,sin\left(\frac{n_0}{n}\right)$$

By repeating the same optical efficiency calculations as those used for the case of air ($n_0 = 1$), we obtain the new expression of η_O for an external medium of refractive index n_0:

$$\eta_O = \frac{1}{4}\left(\frac{n_0}{n}\right)^2\left[1 - \left(\frac{n - n_0}{n + n_0}\right)^2\right] \qquad [1.26]$$

Generally, marketed diodes are wrapped in an epoxy dome to protect the active surface from moisture and mechanical impact. The refractive index of this polymer is $n_0 \sim 1.50$. The optical efficiency of the source with an SC of refractive index $n = 3$, and covered with an epoxy dome, is $\eta_O = 5.50\%$. This value assumes that the light collected in the dome is entirely transmitted to the ambient air. To arrive at this, we place the P–N junction at the center of the hemispherical dome such that all of the rays of light arrive perpendicular to the dome surface and are transmitted without being reflected.

To increase the extraction efficiency of rays emitted, we can *texture the surface* of the SC by employing nanostructures that reflect the incident rays such that they are transmitted to the outside after one or two reflections on the faces of these nanostructures. We can also use distributed *Bragg reflectors*, which are flat, reflective surfaces with different refractive indices. These reflectors, placed between the SC and the substrate, make it possible to reflect the rays emitted towards the output surface, and then towards the outside, with no notable losses.

1.3.1.4. *LED efficiency*

For each physical process occurring in the conversion of electrical energy into optical energy of the LED, a corresponding efficiency can be defined in order to be able to evaluate the effectiveness of each operation.

1.3.1.4.1. Injection of charges and emission of photons

The efficiency of the conversion is evaluated by the *internal quantum efficiency*, which is defined by:

$$\eta_i = \frac{number\ of\ photons\ emitted\ by\ the\ junction\ per\ second}{number\ of\ electrons\ injected\ into\ the\ diode\ per\ second}$$

As a function of the measures, this efficiency is written as:

$$\eta_i = \frac{P_i\,/\,h\nu}{I\,/|e|} \qquad [1.27]$$

In this expression, P_i is the optical power (in W) of the photons emitted by the junction and I is the current intensity or electric current (in A) that flows in the diode.

1.3.1.4.2. Extraction of photons

The efficiency of the photon extraction is evaluated by the *extraction efficiency* or *optical efficiency*, which is defined by:

$$\eta_O = \frac{number\ of\ photons\ emitted\ to\ the\ external\ medium\ per\ second}{number\ of\ electrons\ emitted\ by\ the\ junction\ per\ second}$$

The optical efficiency is given by expression [1.25].

1.3.1.4.3. Injection of charges and emission of photons to the external medium

The efficiency of the conversion of electrical energy into optical energy is evaluated by the *external quantum efficiency*, which is defined by:

$$\eta_e = \frac{number\ of\ photons\ emitted\ to\ the\ external\ medium\ per\ second}{number\ of\ electrons\ injected\ into\ the\ diode\ per\ second}$$

Its expression is as follows:

$$\eta_e = \eta_i \times \eta_O \quad [1.28]$$

Let P_0 be the luminous power of the photons emitted by the diode to the external medium. The number of photons emitted to the external medium per second is equal to $P_0/h\nu$. The expression of the external quantum efficiency is:

$$\eta_e = \frac{P_0/h\nu}{I/|e|} \quad [1.29]$$

1.3.1.4.4. Energy conversion

The efficiency of the energy or power conversion is evaluated by the *energy efficiency* or the *power conversion efficiency* (PCE), which is defined by:

$$\eta_{PCE} = \frac{luminous\ power\ emitted\ to\ the\ external\ medium}{electrical\ power\ supplied\ to\ the\ diode} = \frac{P_0}{I \times V} \quad [1.30]$$

This efficiency has no unit as it is equal to the ratio of two powers (in W). However, in terms of visibility, the luminous power, P_0, is a function of the luminous flux, Φ, as follows:

$$\Phi = P_0 \times K_m \times V(\lambda) \quad [1.31]$$

The parameter K_m is the *maximum photopic spectral luminous efficacy* and is equal to 683 lm/W for the wavelength $\lambda = 555$ nm.

At this wavelength, an optical power of one watt corresponds to a luminous flux of 683 lm.

The function $V(\lambda)$ is the *spectral luminous efficacy function*, representing the sensitivity of the human eye to different colors. It is maximum at the wavelength $\lambda = 555$ nm, which corresponds to the color yellow-green.

The *luminous efficacy* η_{EL} is defined for a lighting source by:

$$\eta_{EL} = \frac{luminous\ flux}{electrical\ power} = \frac{\Phi}{I \times V} \qquad [1.32]$$

It represents the luminous flux measured by an observer with respect to the electrical power used to produce this light. The unit of η_{EL} is the lumen per watt (lm/W).

A 60 W lamp producing a power of 900 lm has a luminous efficiency of 15 lm/W. Note that with regard to lighting, luminous efficiency is often referred to as luminous efficacy of the light source.

1.3.2. *OLEDs*

The basic operating principle of OLEDs is the same as that of inorganic diodes and certain concepts introduced in section 1.3.1 are also applicable to OLEDs.

However, the organic nature of emitters in OLEDs requires the introduction of other parameters, which we examine in this section.

1.3.2.1. *Operating principle*

1.3.2.1.1. Basic OLED structure

In a basic OLED structure, under the application of a voltage V_0 at the diode terminals, the electrons are injected from the cathode and the holes of the anode into the active organic layer. The charge carriers of mobilities μ_n

and μ_p move and can recombine to emit a photon constituting the light emitted by the diode. Electroluminescence processes comprise the following steps:

– injection of charges into the SC from the electrodes;

– transport of charges injected under the influence of the electric field;

– formation of excitons in the SC;

– radiative recombination of excitons and emission of photons;

– extraction of photons into the external medium.

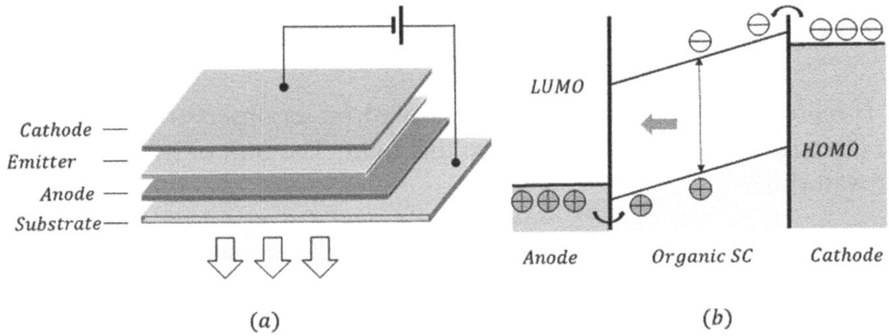

Figure 1.5. *Structure and operating principle of a basic organic light-emitting diode. For a color version of this figure, see www.iste.co.uk/nguyen/electronics2.zip*

The transparent conducting anode is very often a glass or plastic substrate covered with a layer of ITO. The metal cathode is often a thin layer of metal of low work function. The active layer is an organic SC of small molecules or conjugated polymer.

The injection, diffusion and recombination processes are described in Volume 1. Note that not all charge carriers recombine as some can cross the active layer without encountering a carrier of the opposite sign. These carriers constitute the leakage current (of holes or electrons). At the origin of these leakage currents are, first, the presence of electrically active traps in the interface regions and in the active layer that capture the carriers. Second, in most organic SCs, mobilities μ_n and μ_p are different, leading to two charge carrier currents, one majority and one minority. This dissymmetry reduces the probability of the recombination of injected carriers in the active layer. In both cases, the charge carriers in the leakage currents do not contribute to light emission in the OLEDs.

1.3.2.1.2. The p-i-n or n-i-p concept

In a basic OLED structure, the emitting layer plays a double role, assuring both the injection and transport of charge carriers and the emission of photons. These two processes, which occur simultaneously, cannot be optimized to exploit the potentials of the emitting material. To remedy this, the *p-i-n* or *n-i-p concept* has been proposed, according to which charge transport in the emitting material is facilitated by the upstream preparation of the electrode/emitter interface when carriers are injected from the electrodes. We can distinguish between the injection of holes from the anode and that of electrons from the cathode and by inserting transport layers between the electrodes and the emitter, we improve the injection of charges into the emitter.

The elaborate OLED structure comprises in addition an electron transport layer (ETL) on the cathode side and a hole transport layer (HTL) on the anode side. The role and operation of these layers are explained in Chapter 4 of Volume 1. By doping the HTL (like a P-type SC) and ETL (such as an N-type SC) transport layers, we can improve their conductivity and significantly reduce ohmic losses at the contacts. As a result, the diode turn-on voltage becomes lower and leads to a better conversion efficiency. Another advantage of the p-i-n structure lies in the fact that the Fermi level of the metal electrode is similar to that of the transport layer and, if the doping of the latter is substantial, the charge carriers can cross the Schottky barrier through the tunnel effect. Under this condition, the charge injection is almost ohmic and the amount of carriers injected notably increases. We sometimes find other buffer layers, which are added to the structure developed. These layers play, on the one hand, a blocking role for a given carrier type and prevent these carriers from passing to another layer. They are called the *hole blocking layer* (HBL) and the *electron blocking layer* (EBL). On the other hand, they play the role of the transport layer for charge carriers of the opposite sign to the blocked carriers and promote transport of these carriers. Thus, a hole blocking layer is also an electron transport layer, and vice versa. Figure 1.6 presents the energy diagram of an OLED with different transport layers and blocking layers.

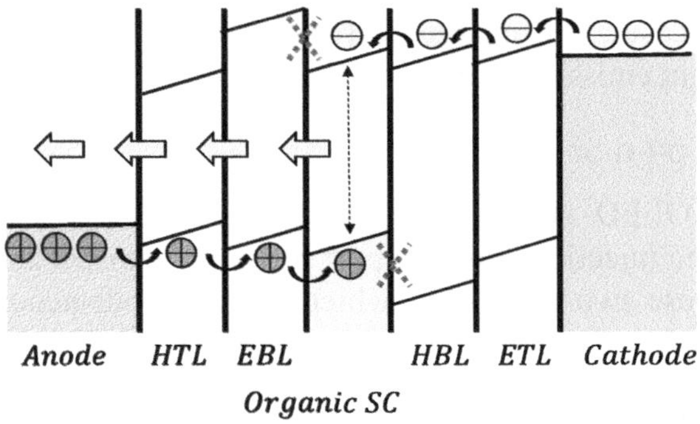

Figure 1.6. *OLED p-i-n structure with hole and electron blocking layers. For a color version of this figure, see www.iste.co.uk/nguyen/electronics2.zip*

Here, the holes cannot pass over the potential barrier between the SC and the HBL, because the height of the barrier is substantial owing to the choice of HBL material. By contrast, the passage of electrons from the ETL to the HBL is promoted by the low value of the potential barrier existing between these two layers; likewise for the SC electrons, which cannot cross the potential barrier between the SC and the EBL. As a result, using HBL and EBL results in the confinement of charge carriers in the SC, and the number of recombinations increases, as does the number of photons emitted.

1.3.2.2. *Exciton formation and recombination*

In organic SCs, excitons are located on the molecules. The spin of the exciton formed is determined from the combination of the spins, M_S, of the electron and the hole. There are two different states for the excitons formed: singlet states of multiplicity $M_S = 0$ and triplet states of multiplicity $M_S = 1$. Due to the indistinguishability of the particles, statistically excitons in the singlet state represent 25% and those in the triplet state represent 75% of the excitons created. In principle, only excitons in the singlet state perform radiative transitions and contribute to the OLED light emission. This distinction between the exciton states leads to the theoretical limitation of useful recombination for OLEDs being 25%.

1.3.2.2.1. Exciton recombination in the singlet state

Recombination of an exciton occurs when an electron and a hole come together at a critical distance, called the *Onsager radius*, for which the

Coulomb interaction is greater than the thermal agitation. In an SC of permittivity $\varepsilon = \varepsilon_0 \varepsilon_r$, the Onsager radius is:

$$r_c = \frac{|e|^2}{4 \pi \varepsilon k T} \quad [1.33]$$

The exciton recombination rate as a function of the concentrations of electrons and holes is given by the relation [1.9]:

$$R = B n p$$

The bimolecular recombination coefficient, B, as a function of the Onsager radius, r_c, is given by the following relation:

$$B = 4 \pi r_C (D_n + D_p), \quad [1.34]$$

where D_n and D_p are the diffusion coefficients of electrons and holes in the SC. By combining relations [1.33] and [1.34] with the Einstein relation, we obtain the following expression of the recombination rate:

$$R = B n p = \frac{|e|}{\varepsilon} (\mu_n + \mu_p) n p \quad [1.35]$$

As we have mentioned, the mobilities of electrons and holes are very different in organic SCs. The recombination rate is higher for SCs with more mobile carriers and a high concentration of these carriers. We preferably choose P-type organic materials for the emitters. In reality, when the carrier mobilities show a large difference, the OLED efficiency is low as the recombination does not occur in the organic SC but in the vicinity of the electrode where low-mobility carriers are injected. In this case, radiative recombination risks are reduced by the interface states and the defects, which are numerous in this region. To obtain optimum OLED efficiency, it is preferable for the mobilities of holes and electrons to be similar in order to ensure their recombination in the volume of the organic SC. In this case, we say that there is a *balance of charges* transported or charge balance.

1.3.2.2.2. Exciton recombination in the triplet state

The relaxation of a molecule from a triplet state to the ground state is generally non-radiative. The radiative transitions from this state, if any, correspond to phosphorescent emission, the efficiency of which is lower than that of the fluorescent emission of the same molecule. Indeed, the relaxation time of the triplet state is long and facilitates the occurrence of non-radiative processes (emission of phonons) that alter light emission efficiency. Statistically, of the total number of excitons created, excitons in the triplet state represent 75%, compared to 25% in the singlet state. It is therefore interesting to identify the means to exploit excitons in the triplet state and thus increase the emission of photons of active material. One of them consists of doping the fluorescent material with phosphorescent molecules to obtain the emission of two materials. The emission process can be described as follows: assuming that the fluorescent and phosphorescent materials are compatible, an interaction between the triplet states is produced to begin with, called *triplet–triplet annihilation* (TTA) (Kondakov 2015), then a process of exciton transfer from triplet states to singlet states (called *inverted intersystem crossing*) takes place according to:

$$T_1 + T_1 \rightarrow S_0 + S_1 \qquad [1.36]$$

The exciton in the singlet state relaxes to the ground state by emitting a photon, and the emission (called *delayed fluorescence*) is added to the original emission of the emitting material and thus increases the internal quantum efficiency.

To make the delayed fluorescence process possible, the materials need to be compatible, that is, the following conditions need to be met:

– the lifetime of triplet states must be long and that of singlet states short;

– the difference in energy between the singlet, S_1, and triplet, T_1, states must be very low (< 100 meV) (Uoyama *et al.* 2012). The process of exciton transfer by inverted intersystem crossing can be thermally activated and therefore significantly increases the efficiency of the diodes. The emission of photons is called *thermally activated delayed fluorescence* (TADF).

The best results are obtained with phosphorescent materials composed of metal or organometallic complexes of heavy transition metals such as Pt (platinum), Re (rhenium), Os (osmium) and Ir (iridium).

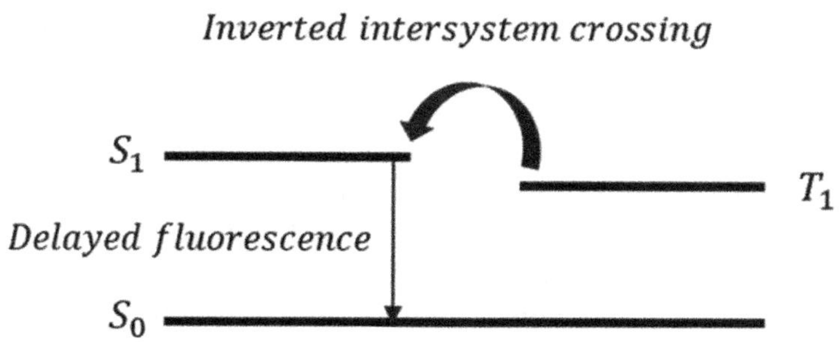

Figure 1.7. *Inverted fluorescence process*

1.3.2.3. *Materials and colors*

1.3.2.3.1. Fluorescent materials

Many organic materials exist that emit light of various colors in the visible range, with the exception of the color white. These materials comprise small molecules and polymers. Figure 1.8 gives a few examples of organic emitting materials with the emission colors.

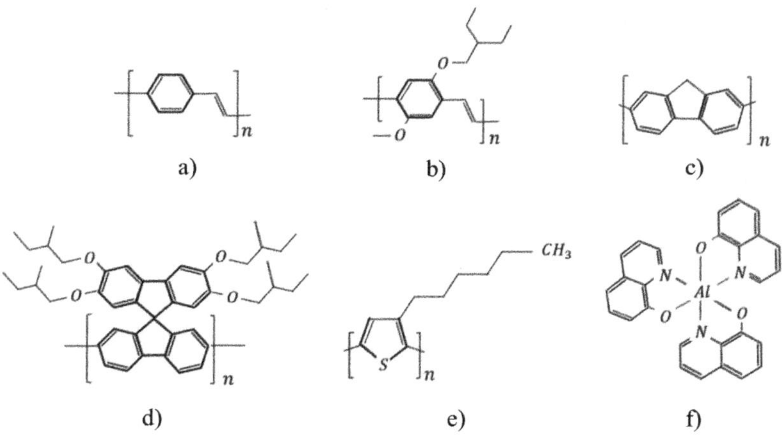

Figure 1.8. *Main fluorescent materials: a) poly(phenylene-vinylene) (PPV) – green; b) poly[2-methoxy-5-(2-ethylhexyloxy)-1,4-phenylene-vinylene] (MEH-PPV) – red; c) polyfluorene (PF) – blue; d) poly(spirobifluorene) (PSBF) – blue; e) poly(hexylthiophene) (P3HT) – red; f) poly[2-methoxy-5-(2-ethylhexyloxy)-1,4-phenylene-vinylene] (Alq3) – green*

By modifying the chemical structure of an emitting conjugated polymer, it is possible to control the color of its emission. For example, PPV is an emitting polymer of yellow-green light. By adding electron donor groups, for example alkoxy groups, $R - O$, to the phenyl rings in positions 2 and 5, the polymer emission turns red. It is also possible to modify the polymer gap value by reducing its conjugation length, which results in a blue shift of the emission spectrum. This solution consists of introducing, in a controlled manner, non-π-conjugated units or groups into the backbone of the polymer, thus reducing the conjugation length of the initial polymer. Certain other conjugated polymers emit blue light, such as poly(p-phenylene) (PPP) or poly(alkylfluorene) (PF). The fluorescent emitting materials most widely used to create OLEDs are shown in Figure 1.8.

1.3.2.3.2. Phosphorescent materials

Of the phosphorescent materials used in OLEDs, platinum and iridium possess good properties for heavy transition metal complexes. The iridium complex, *Irppy*3, is incorporated into the diodes and significantly increases the emission efficiency thanks to a very short relaxation time compared to that of platinum complexes. Figure 1.9 gives a few examples of organic emitting phosphorescent materials with the emission colors.

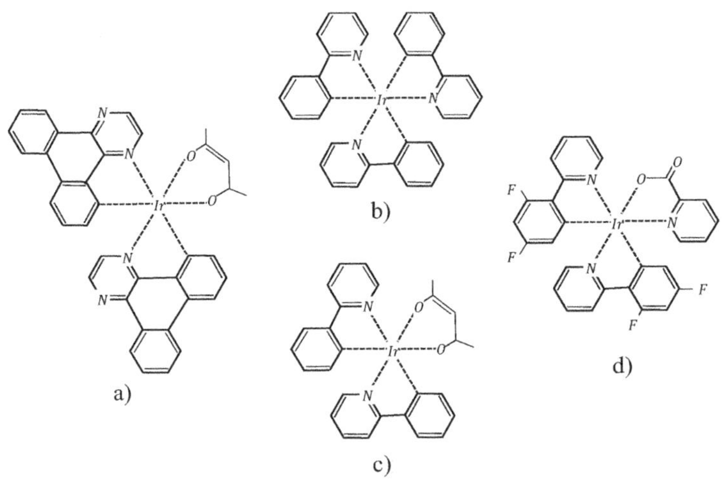

Figure 1.9. *Main phosphorescent materials: a) Ir(MDQ)2(acac) – green; b) Ir(ppy)3 – red; c) Ir(ppy)3(acac) – red; d) FIrpic – blue*

1.3.2.3.3. Doping

In order to modify the emission color of a material, a small quantity of dopants is added to a matrix of emitting material. As we already saw in Chapter 2 of Volume 1, the doping of organic materials consists of replacing certain molecules in the matrix with other molecules to exploit the optical and electronic processes provided by the doping molecules and thus improve the material's performance.

An emitting material can be indifferently doped with fluorescent or phosphorescent dopants. *Fluorescent dopants* are generally dyes whose emission corresponds to the primary colors (RGB). When added to an emitting host matrix, the transfer of energy between molecules allows the formation of excitons on the dopant and results in the emission of light of the color of the dye, with high efficiency. *Phosphorescent dopants*, generally transition metal-based complexes, are added to a fluorescent organic host matrix. The triplet states of fluorescent material are not photon emitters and the triplet states of phosphorescent material present low emission efficiency. Doping makes it possible to use the singlet and triplet states of the two materials in association to increase light emission. The process can be explained as follows: singlet and triplet excitons created in the matrix material are transferred to the triplet states of the phosphorescent dopant. By inverted intersystem conversion, they are then transferred to the singlet states of the dopant, which de-excite by emitting photons. The interaction between the host matrix and the guest dopant occurs by Förster or Dexter energy transfer between molecules of the two materials.

Dexter transfer occurs from the matrix to the dopant when an electron passes from a singlet state or triplet state of the matrix to a corresponding state of the dopant. At the same time, an electron from the ground state of the dopant changes to the ground state of the matrix. This transfer conserves the multiplicity of spin and it is possible between the singlet states (singlet–singlet) or the triplet states (triplet–triplet). It requires an overlap of molecules, which must be close to one another. In other words, the doping level needs to be high.

Förster transfer occurs from the matrix to the dopant by transferring its energy to it. The molecule of the matrix is in the ground state after the transfer, whereas the dopant molecule becomes excited. This transfer takes place by emission and does not require high doping levels.

1.3.2.4. *OLED and white-light emission*

As in LEDs, white-light emission by OLEDs is obtained by combining light emissions of different wavelengths. Mainly, the mixture of the primary colors RGB is the most widely used, but we can also obtain the white color with a mixture of blue and yellow colors.

The implementation techniques and the structure of organic diodes are, however, different from those used for LEDs. The easy preparation of mixtures of organic materials with particular physical properties facilitates the implementation of devices on any type of substrate.

1.3.2.4.1. Red-light emission

Red-light emitting diodes use emitting materials as a doped matrix with red-light emitting chromophores. To ensure an efficient energy transfer between the two materials, the emission and absorption spectra of the two materials should overlap. Alq_3 and 2-tert-butyl-9,10-di(naphtha -2-yl)anthracene (TBADN) are commonly used as the host material. Red chromophores such as 4-(dicyanomethylene)-2-t-butyl-6-(1,1,7,7-tetramethyljulolidyl-9-enyl)-4H-pyran (DCJTB) or the derivative 4-(dicyanomethylene)-2-i-propyl-6-(1,1,7,7-tetramethyljulolidyl-9-enyl)-4H-pyran (DCJTI) are used as dopants. The luminous efficacy of red OLEDs remains low (~ 1 lm/W).

1.3.2.4.2. Green-light emission

Green-light emitting dopants are among the most efficient materials used to obtain color OLEDs. Chromophore C-545T or 10-(2-benzo-thiazolyl)-1,1,7,7-tetramethyl-2,3,6,7-tetrahydro-1H,5H,11H-[1]benzo-pyrano[6,7,8-ij] quinolizin-11 and its derivatives are widely used for doping Alq_3 matrices, and the efficiency of the diodes is particularly high thanks to the very low rate of fluorescence quenching in these materials. Its most widely used derivatives are C-545MT (addition of a methyl group at the C-4 position) and C-545TB (substitution of a butyl group at the benzene ring).

1.3.2.4.3. Blue-light emission

Blue-light emitting dopants with good emission efficiency are difficult to obtain. The efficiency obtained with the blue-emitting OLEDs varies between 6 and 11 cd/A, which is lower than that of the red-light emission (8 – 11 cd/A) or the green-light emission (15 – 20 cd/A) (Ricks *et al.* 2007). In addition, blue-emitting diodes are unstable and their lifetime is generally shorter than that of the green or red emitters. The polymers and dopants often used for blue-light emission are 4,4-bis(2,2-diphenylvinyl)biphenyl (DPVBi)/4,4'-Bis(9-ethyl-3-carbazovinylene)-1,1'-biphenyl (BCzVBi) or 2,5,8,11-tetra-tert-butylperylene (TBP)/perylene.

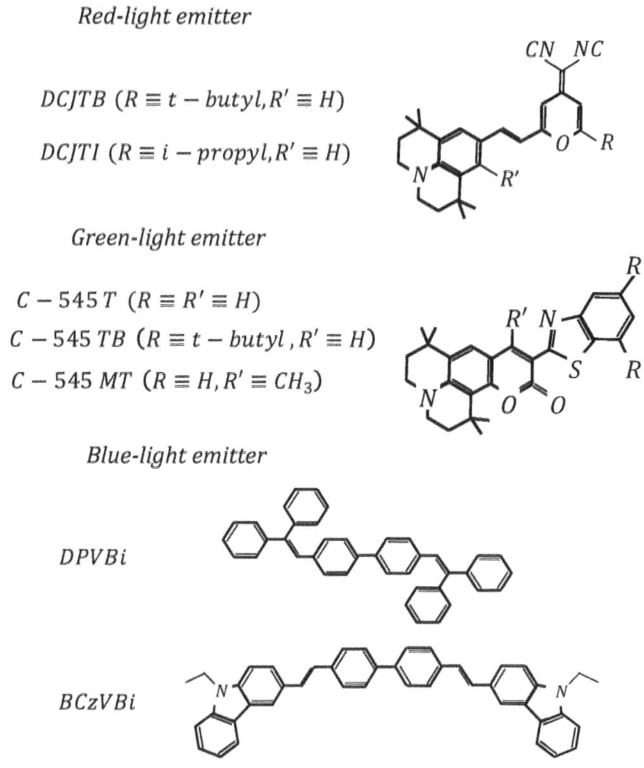

Figure 1.10. *RGB light emitters*

1.3.2.4.4. White-light emission

White-light emission by OLEDs can be obtained with the same techniques as those used for LEDs. Moreover, thanks to the small thicknesses of the layers used, new structures can be designed to produce white diodes. Likewise, doping by grafting light-emitting groups onto the polymer backbone enables the production of white diodes from a single organic material.

The following techniques are applied to obtain white light in OLEDs:

– A mixture of the primary colors, RGB, by three emitting layers, vertically stacked or in laterally patterned pixels.

– A mixture of two or more polymers, whose emission spectra cover different ranges of the visible spectrum. The resulting spectrum broadly corresponds to the visible spectrum and the light emitted is white. For example, by mixing polyfluorene (PFO, blue emitter) with MEH-PPV (red emitter) to produce the active layer of an OLED, the luminous efficiency obtained is 12.6 lm/W (Huang *et al.* 2006). It is also possible to use a single polymer as the blue-emitting material and modify its emission color (to green or red) by substituting the functional groups on the backbone of the chain. Then, by mixing the polymers emitting RGB colors, a white-emitting active layer is obtained. For example, the PPF-SO polymer based on 9,9-bis(4-(2-ethylhexyloxy) phenyl)fluorene (PPF) containing a dibenzothiophene-S,S-dioxide (SO) group is a blue emitter. The emission color of this polymer is changed to green with the addition of a group of benzothiadiazole (BT) (abbreviated as PPF-SO-BT) and to red with a group of 4,7-di(4-hexylthien-2-yl)-benzothiadiazole (DHTBT) (abbreviated as PPF-SO-DHTBT). The copolymer formed from PPF-SO, PPF-SO-BT and PPF-SO-DHTBT emits white color with good stability and a luminous efficacy of 5.5 lm/W (Yu *et al.* 2013). Certain phosphorescent materials, added to the host polymer, are used to obtain white-light emission. They increase efficiency and strengthen emission stability. For example, tris[2,5-bis-2'(9',9'-dihexylfluorene)pyridine]iridium(III) (Ir(HFP)) red emitter is added to the copolymer formed from poly(9,9 dioctylfluorene)-cofluorenone) (PFO-F) blue and green emitter to form the active layer of a white diode of color temperature ~ 4,600 K and color rendering index ~ 86, which are very close to the characteristics of visible light (Gong *et al.* 2004).

– Doping of a blue-emitting polymer by dopants emitting light whose spectrum is complementary to that of the host material. The resulting emission spectrum covers the entire visible range and the light emitted is white. The main difficulty lies in the control of the energy transfer mechanism between the host material and the guest dopants. Energy transfer between a wide-gap material and a narrow-gap material may occur by Förster or Dexter mechanisms. The dopants used can be phosphorescent polymers or quantum dots (QDs). For example, a white-light emitting OLED is made using a copolymer of poly[(9,9-dihexyloxyfluoren-2,7-diyl)-alt-co-(2-methoxy-5-(2-ethylhexyloxy)phenylen-1,4-diyl)] (PFH-MEH) doped with QDs of cadmium selenide (CdSe) (Li *et al.* 2005). The structure of the diode is ITO/PEDOT:PSS/ (PFH-MEH:CdSe/ZnS/Alq_3/Ca/Al). The Alq_3 layer serves as a transport layer and a green-light emitting layer. QDs are 5 nm in size and emit red light by energy transfer from the PFH-MEH copolymer, which emits blue light, and from Alq_3, which emits green light.

– Other techniques: the emission of white light can be obtained using the emission of phosphorescent excimers associated with that of the fluorescent material. The white-light generation mechanism can be summarized according to the following diagram: a transfer of charges between the electron donor material *(D)* and the electron acceptor material *(A)* creates exciplexes. The emission spectrum of the exciplexes covers a wide range of wavelengths localized between that of the electron acceptor material *(A)* and that of the electron donor material (*D*) (Kalinowski *et al.* 2007). Three radiative states are active at the same time: the triplet states, $^3A*$, of the phosphorescent monomer; the excimers in the triplet state, $^3(AA)*$; and exciplexes, $^3(DA)*$. The resulting spectrum is generally that of visible light with a low luminous efficacy.

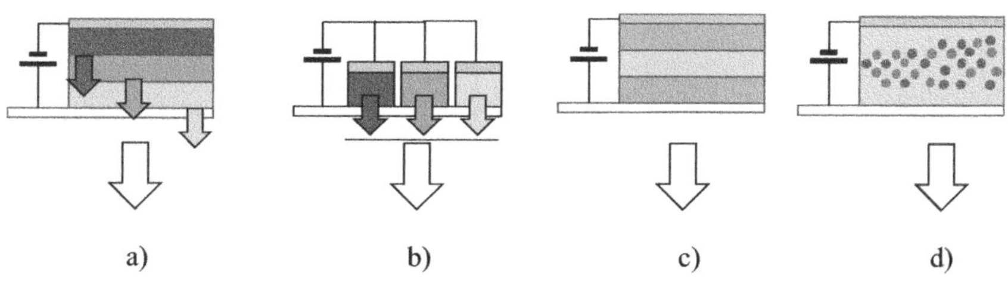

Figure 1.11. *Emission of white light: a) by mixing the colors of stacked RGB-emitting layers; b) by mixing the colors of the emitting layers in pixels; c) by mixing the colors of the emitting layer and the transport layers; d) by mixing the colors of the emitting layer and the dopants (down-conversion process). For a color version of this figure, see www.iste.co.uk/nguyen/electronics2.zip*

1.3.2.5. *Emission of light to the external medium*

As in LEDs, light emission from an OLED to the external medium is strongly influenced by the structure of the diode. The problems of refraction at the SC/external medium interface encountered in inorganic LEDs apply also to OLEDs. There are, however, two notable differences between organic and inorganic diodes. First of all, the emitting layer of the OLED is not in direct contact with the ambient air but with the transparent electrode, in this case ITO, which generally covers the glass substrate. The light emitted passes through a minimum of two layers before exiting into the ambient air. In addition, the layers of an OLED are thin films with an average thickness of a few tens of nanometers. From an optical point of view, the diode structure can be considered as a *microcavity* formed by the emitting layer placed between the transparent electrode and the metal electrode. Certain light rays emitted by this cavity are reflected at the interfaces and confined within the volume of the microcavity. The spectrum of light emitted into the external medium varies as a function of the thickness of the cavity, that is, the thickness of the films. This process is called the microcavity *interference effect*. Consequently, the use of certain techniques already seen in LEDs becomes problematic for OLEDs as they require strict, precise control of the deposition of additional layers (Bragg reflectors, for example).

1.3.2.5.1. Optical efficiency

Figure 1.12 shows the loss mechanisms during the extraction of light to the external medium in a simple OLED structure comprising an active layer inserted between an anode and cathode.

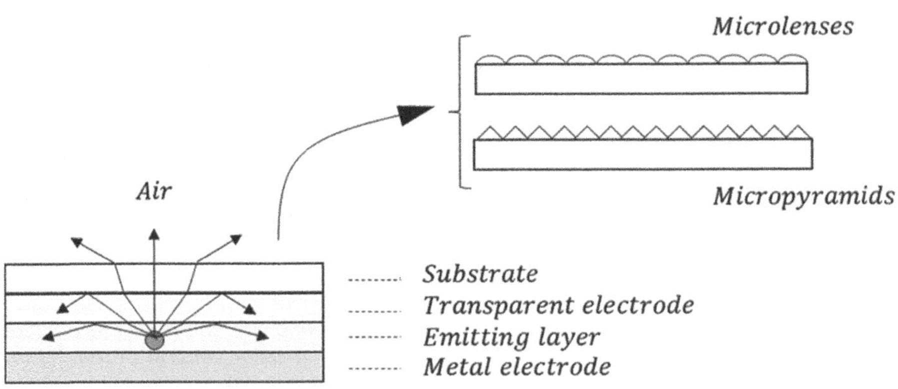

Figure 1.12. *Optical losses during extraction of light emitted towards the outside in a simple OLED structure. For a color version of this figure, see www.iste.co.uk/nguyen/electronics2.zip*

Total reflections may take place at the interfaces between the layers and only a fraction of the radiations emitted will exit into the air. This fraction is used to define the optical efficiency of the diode.

The *extraction efficiency or the optical efficiency* is obtained by integrating the intensity emitted on the surface of the extraction cone, taking into account the multilayer structure of the diode, the reflection of the radiations on the interfaces, the emission on the edges and other minor factors that affect emission (Tsutsui and Takada 2013). Let n be the refractive index of the emitting material. The optical efficiency is given by (Kim *et al.* 2000):

$$\eta_O = 1 - \sqrt{\left(1 - \frac{1}{n^2}\right)} \qquad [1.37]$$

If $n > 1.6$, the value of the optical efficiency is approximatively equal to $\eta_O \sim 0.5\, n^{-2}$. As the refractive index of the organic materials composed of aromatic molecules is close to 1.60, the optical efficiency of these materials is often estimated at 20%.

1.3.2.5.2. Improving optical efficiency

We can identify two optical loss sources in OLEDs. The first source concerns the extraction of light at the substrate/air interface, which can be qualified as external losses or substrate mode. The second loss source concerns the total reflection at the organic layers and the electrode layers, qualified as internal losses or internal mode.

The technique generally used to reduce external optical losses is to employ arrays of *microlenses* or *micropyramids* on the surface of the substrate to facilitate the refraction of radiations emitted and increase the extraction efficiency (see Figure 1.12). These arrays play the role of the epoxy dome covering the inorganic SC emitting surface and enabling the transmission without reflection of rays emitted to the external medium. The diameter of the microlenses is approximately 10 μm. Improvement of the optical extraction at the substrate is nevertheless limited in order to increase the efficiency of the diode, as internal losses cannot be avoided.

For reducing internal optical losses, a technique has been proposed to facilitate the diffusion of the radiations by interposing a diffusing layer of low refractive index ($n \sim 1.03$) between the substrate and the organic layer (Tsutsui *et al.* 2001). This layer deflects radiations normally trapped in the structure of the diode, enabling them to escape into the external medium. It can consist of a microgrid inserted between the substrate and the active layer (Sun and Forrest 2008) or a transparent ITO electrode shaped into a microstructure (Koh *et al.* 2010). These techniques are efficient but also have drawbacks, such as an increase in electrical resistance, which tends to degrade the charge transport. Using a polymer/nanoparticle composite to obtain a thin film acting as a light diffuser has been shown to substantially increase diode luminous efficacy. For example, using a composite film of propylene glycol-monomethyl-ether acetate (PGMEA)/nanoparticles of TiO_2 has made it possible to increase the external quantum efficiency of a phosphorescent OLED by 50% compared to that of a similar diode without the layer of low refractive index (Chang *et al.* 2013).

Other techniques can be used to increase the extraction efficiency of light emitted by the active layer of OLEDs. For example, the use of substrates with a high refractive index (of a value close to that of the emitting organic material) would enable the light emitted to cross the organic/substrate interface without reflection, and therefore be transmitted to the outside with a high efficiency (Nakamura *et al.* 2005). However, these techniques do not at present offer a satisfactory solution, as they are costly to implement or significantly modify the structure of the diode and alter its optoelectronic properties.

1.3.2.5.3. Inverted-structure OLED

Extraction of light from OLEDs can be facilitated if the photons are reflected by the lower metal electrode and exit the diode into the air through the upper transparent electrode (generally ITO), which has a high transmission coefficient. Such structures are called TOLEDs (top-emitting organic light-emitting diodes) and different diode configurations can be obtained (see Figure 1.13).

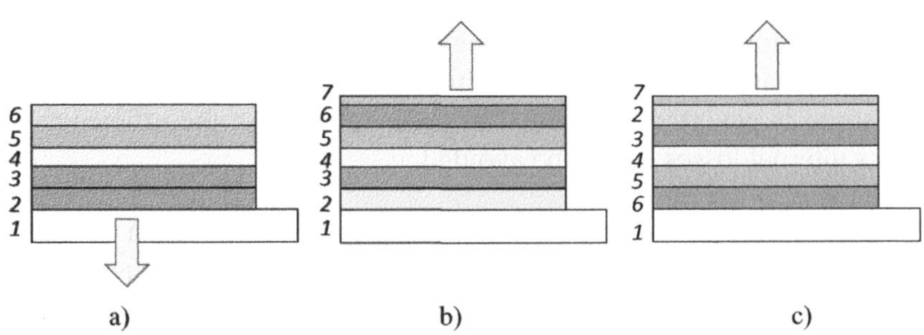

Figure 1.13. *The diode structure of: a) a standard OLED; b) a standard TOLED; c) an inverted TOLED. The different layers are: (1) substrate; (2) anode; (3) ETL; (4) emitter; (5) HTL; (6) cathode; (7) dielectric layer. For a color version of this figure, see www.iste.co.uk/nguyen/electronics2.zip*

There are two types of TOLED. Normal TOLEDs use semi-transparent metallic layers to replace ITO while retaining the functional structure of the standard OLED. Inverted TOLEDs are obtained by inverting the layer deposition sequence. The anode of these TOLEDs is composed of ITO or semi-transparent metal. Compared with the standard structure, this structure enables the refraction of light rays to be suppressed at the ITO/substrate interface and the optical coupling is improved. In addition, with the transparent electrode deposited on the organic layer, the interaction between the oxide and the organic film can be low or negligible. Indeed, in a standard structure, the deposition of polymer layers in solution on the ITO can react with the oxide to release metal atoms, which will diffuse in the active layer and alter the physical processes. It has in fact been observed that the diffusion of metal atoms from ITO occurs in both cases of ITO/active layer and active layer/ITO formation (Nguyen *et al.* 2006). The use of a buffer layer (transporting and/or blocking) between the ITO and the active layer is necessary to avoid the interaction between the two materials. The use of *aluminum-doped zinc oxide* (AZO) instead of ITO for transparent electrodes is an interesting alternative because it is less expensive than ITO and does not require the use of rare metals (indium). Moreover, the deposition of a dielectric layer on top of the upper electrode (the cathode for standard TOLEDs and the anode for inverted TOLEDs) improves the extraction of light towards the external medium and increases efficiency (Riel *et al.* 2003).

1.3.2.6. *Transparent OLED*

In certain applications, it is beneficial for the user to have a display screen, which becomes transparent when switched off. They will then be able to see the objects placed behind the screen as if looking through a simple pane of glass. These display screens cannot be made using inorganic LEDs, but it is possible to modify the structure of OLEDs to make them transparent when they are switched off. Figure 1.14 shows the basic structure of a transparent OLED.

In these diodes, the electrodes are transparent conducting oxides (TCO). Like TOLEDs, the active layer must be separated from the electrodes to avoid interactions between materials, leading to the diffusion of metal atoms in the emitter, and consequently causing unstable operation of the diode. The buffer layers inserted should be sufficiently transparent and should facilitate the injection of charges into the emitter. In addition, their deposition should not damage the active layer. It is difficult to find materials possessing all of the physical properties required and often several layers need to be combined to create a buffer layer.

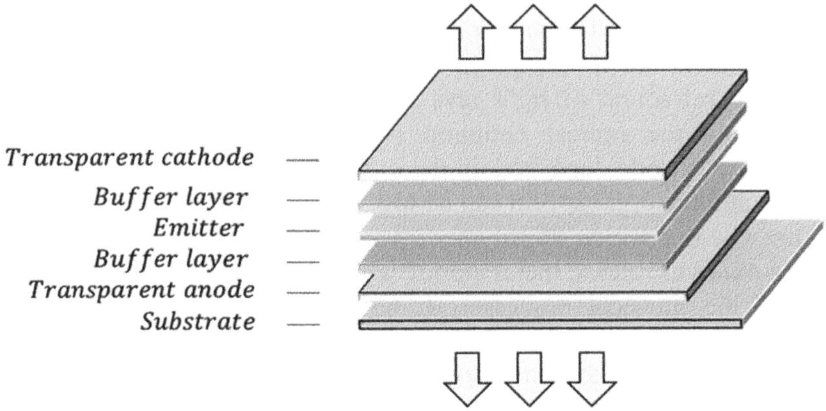

Figure 1.14. *Structure of a transparent OLED. For a color version of this figure, see www.iste.co.uk/nguyen/electronics2.zip*

On the anode side, transition metal oxides such as MoO_3, WO_3 and NiO are transparent and frequently used as the hole injection material and buffer layer. On the cathode side, different materials (e.g. MgAg (Gu *et al.* 1996), lithium-doped Bphen (Zhou *et al.* 2002)) are used as the electron injection material and the buffer layer of low thickness (~10 nm) with good transparency for light transmission.

1.3.2.7. *OLETs*

Organic light-emitting transistors (OLETs) combine the light-emitting properties of OLEDs and the switching properties of field-effect transistors to create applications that can come to be of interest. *Organic field-effect transistors* (OFETs) are studied in Chapter 3.

The schematic diagram of an OLET is given in Figure 1.15.

The basic structure of an OLET consists of three electrodes: the source, *S*, the drain, *D*, and the gate, *G*, which are arranged in the device in the same way as a conventional field transistor. The electrons (holes) are injected from the source into the emitting layer when the transistor is positively (negatively) biased. The source and drain electrodes are deposited on the surface of the emitting layer and a dielectric layer (generally an insulating, transparent non-conjugated polymer material) is deposited between the emitter and the substrate. The applied voltage at the source and drain electrode terminals is the bias voltage of the LED, which has a planar structure. The voltage applied to the gate electrode enables modulation of the electric field in the channel created in an emitting layer, which, in turn, controls the intensity of the current and therefore the intensity of the emission of light to the exterior (through the transparent substrate or the portion of the emitting layer between the source and the drain).

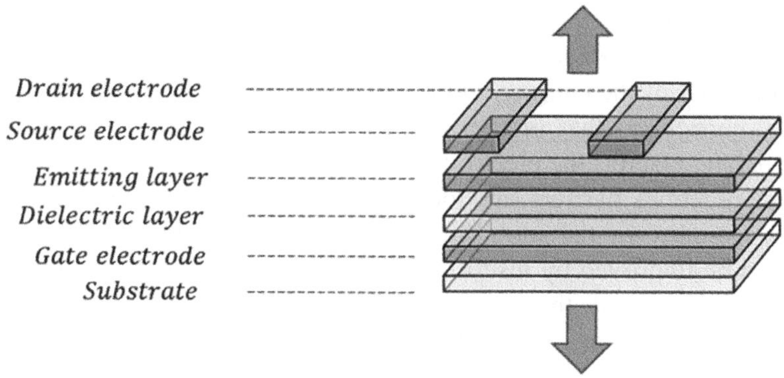

Figure 1.15. *Structure of an OLET. For a color version of this figure, see www.iste.co.uk/nguyen/electronics2.zip*

In the more elaborate OLET structures, ETL and/or HTL transport layers are used as they are for an OLED, to improve the charge transport and recombination of the excitons. These transport layers can potentially emit light and their emission is added to that of the emitter to obtain bi- or tricolor OLETs (Capelli *et al.* 2010). The performance of the OLET depends largely on the quality of the dielectric layer, which is often poly(methyl-methacrylate) (PMMA). This insulating layer plays a significant role in controlling the conductive channel current and influences the operating voltage through its often high capacitance. For this voltage to be reasonably compatible with OLEDs, very thin PMMA films free of any notable defects need to be produced, which requires a high degree of control of the film deposition conditions using solutions. The prospects for the application of OLETs lie in the manufacture of multicolor diodes using multilayer structures but their production would require complex manufacturing processes, most likely at a high cost.

1.3.2.8. *OLED efficiency*

In terms of efficiency, the parameters introduced in section 1.3.1.4 concerning the efficiencies of LEDs are also applicable to OLEDs. With the physical processes being different in OLEDs, we introduce other efficiency definitions, which partly relate these different processes.

1.3.2.8.1. Charge balance factor

The charge balance factor, γ, is equal to the ratio of the number of electrons (holes) to the number of holes (electrons) injected from the electrodes that recombine to create excitons in the active material. It depends on the mobility of the charges and is equal to the ratio of the concentration of minority carriers to that of the majority carriers. In order to optimize the charge balance factor, the electrode materials need to be chosen such that the injection potential barriers are identical or very close to one another. By contrast, equality of the mobilities of the carriers is rarely obtained. This factor can be improved by inserting blocking layers of holes and/or electrons into the structure of the diode, or by modifying the composition and thickness of the electron or hole transport layers.

1.3.2.8.2. Singlet/triplet ratio

Two types of excitons are formed by recombining injected electrons and holes: singlets and triplets. In the fluorescence process, the relaxation of the charges of singlet states is radiative, whereas that of the triplet states is non-radiative. As the spin states are equiprobable, there are three possibilities for forming a triplet exciton and only one for forming a singlet exciton, the singlet/triplet ratio is equal to 1/3 and the probability, β, of fluorescent emission is limited to 25%. By contrast, this probability is 100% for phosphorescent emitters because in these materials the singlet and triplet excitons can emit light.

1.3.2.8.3. Luminescence efficiency

The relaxation of excited states to the ground state can be radiative or non-radiative. Photoluminescence efficiency, η_{PL}, is defined as the ratio of radiative recombination to non-radiative recombination of excitons. The presence of traps in the material promotes non-radiative recombination by capturing the carriers during electronic transitions. These defects can be created by the impurities introduced into the material during deposition of the films or diffused from the electrodes to the organic layer after their formation. The contact region between the electrodes and the organic layers is particularly important for recombination as the energy transfer process between the metal and the organic material is promoted by the formation of dipoles. In this region, the recombination is generally non-radiative.

1.3.2.8.4. OLED efficiency

Taking account of the charge balance factors, spin states and photoluminescence efficiency, the expression of the internal quantum efficiency of an OLED is written as:

$$\eta_i = \gamma \times \beta \times \eta_{PL} \qquad [1.38]$$

The external quantum efficiency (see expression [1.28]) becomes:

$$\eta_e = \gamma \times \beta \times \eta_{PL} \times \eta_0 \qquad [1.39]$$

For a fluorescent-polymer OLED, assuming that the charge balance efficiency is 100%, the singlet exciton generation efficiency is 25%, the photoluminescence efficiency is 100% and the optical coupling efficiency is 20% (with a refractive index of ~1.6), we obtain an external quantum efficiency of $\eta_e = 1\times0.25\times1\times0.2 = 0.05$ or 5%. This efficiency is improved to ~25% in green-emitting OLEDs using combinations of fluorescent and phosphorescent emissions with the TADF process (Cho *et al.* 2014).

1.4. OLED applications

The first OLED applications appeared in the form of simple display screens, replacing inorganic LEDs. These screens were unreliable and the products marketed were quickly withdrawn from the market. The first OLED-screen television was launched by Sony in 2008. After a brief appearance on the market, its production ceased. At present, several large OLED-screen televisions have been proposed by big-brand manufacturers such as Samsung and LG and even rollable OLED-screen televisions have been marketed. Meanwhile, OLED lamps for urban, household lighting and vehicle equipment are beginning to appear on the market.

1.4.1. *OLEDs for lighting*

Compared to LEDs, OLEDs present several advantages when used as light sources for lighting. First of all, an LED is a point source of light and must be incorporated into an assembly to make up a light fixture. In contrast, an OLED is an autonomous emitting surface, which can be deposited on a substrate of various shapes and materials. Next, an operating LED gives off heat that must be removed to prevent premature degradation of the junction. An OLED gives off very little heat and its operating temperature is low (~ 40°C). However, in order to be used in lighting, OLEDs need to meet a number of requirements: luminance and luminous efficacy, lifetime and cost of use.

For lighting, the luminance of the source needs to be sufficient. It needs to be approximately 1,000 cd/m^2 for decorative lamps and greater than 4,000 cd/m^2 for ceiling lamps or lounge lamps. These high values imply high luminous efficacy of the OLEDs with reasonable panel dimensions. With a luminous efficacy of the order of 100 lm/W and a luminance of 1,000 cd/m^2 (Wu *et al.* 2017), the emitted flux is suitable for a white OLED tile of standard dimensions, used as a lighting source. On the other hand, the lifetime of lighting lamps must be substantial for daily use, of the order of 10,000 hours according to the commonly accepted industry standard. The lifetime of an OLED is defined as the operating time after which the luminance is equal to T% of its initial value. The rate, T%, is generally equal to 50%, but according to certain standards, it is also set at 70%. The lifetime and luminance of a diode are closely linked. Indeed, the diode current intensity, I_d, and the lifetime, T_L, experimentally verify the following relation (Tyan 2011):

$$I_d^n \times T_L = \text{Constant} \qquad [1.40]$$

The exponent, n, is dependent on the structure of the diode used and is typically ~ 1.5. As the diode current intensity is proportional to luminance, the lifetime is also dependent on the luminance. According to expression [1.40], an OLED operating at high luminance degrades faster than when operating at lower luminance. In other words, to extend the lifetime of an OLED, it is better to use small tiles to obtain satisfactory luminance. White OLEDs made from phosphorescent and fluorescent materials have been produced with high performance and are suitable for use in the field of lighting. For a tile of dimensions 8 × 8 cm^2, the luminous efficacy of the diodes is 30 lm/W, the luminance is 3,000 cd/m^2, the rendering index is greater than 90 and the lifetime is greater than 10,000 hours. When phosphorescent materials alone are used to obtain the RGB emissions, this performance is remarkably improved to achieve an efficacy of 87 lm/W and a lifetime of over 100,000 hours (Komoda *et al.* 2012).

It can be noted that the technical criteria for comfortable lighting can be satisfactorily achieved with white OLEDs. The fact remains that the marketing of diodes is only possible if the production cost is reasonable. Furthermore, for the lighting of industrial, commercial and office buildings, the esthetic aspect or the styling of the lamps is not essential but the costs of purchasing, installing and maintaining the light sources determine their choice. On the other hand, as a new technology in the field of lighting, OLEDs will only be adopted if they bring a reduction in the cost of use compared to other technologies. Although the material cost is low (the quantity of materials used in a diode is small) and the operating cost is also low (low power consumption), the overall cost of using an OLED lamp remains high, mainly because of the products' low level of industrialization and marketing. The cost of the light fixture, defined as the cost of the lamp divided by the optical power, estimated for one kilolumen for an OLED lamp of luminous efficiency of 120 lm/W is approximately 7 dollars, or 6 euros, to which are added the electricity consumption costs (Duggan 2005). The cost of OLED lamps at present is higher than this estimate and their performance needs to be improved to reach the standard of lamps on the market. It will be necessary to develop the mass production technologies of the devices in order to reduce the manufacturing cost and make OLEDs competitive with other light sources available on the market.

1.4.2. *OLEDs for display*

In the display field, OLED technology occupies an important place on the market for screens of portable devices (telephones and computers) and televisions. There are two different technologies: *passive-matrix OLED* (PMOLED) *screens* and *active-matrix OLED* (AMOLED) *screens*. Matrices are composed of pixels, each of them being a light source composed of three RGB sub-pixels. Overlaying the sub-pixel emissions produces the desired color point in the CIE diagram. OLED screens are gradually replacing *liquid-crystal display* (LCD) *screens*, whose image display principle is based on appropriate blocking out of the light emitted by a backlight source. As a result, an LCD screen is a passive device because it does not emit light while an OLED screen is active. The optical quality of images displayed by OLED screens is better than that of LCD screens. In addition, the image refresh rate is higher for OLED technology than for LCD technology, allowing images to be displayed more fluidly.

1.4.2.1. *PMOLED: passive-matrix screens*

A PMOLED screen consists of a matrix composed of rows (forming the cathode of the diode) and columns (forming its anode) deposited on a substrate. The intersection of a column and a row constitutes an elementary pixel (or an OLED). Each column is powered by a current source that delivers current pulses to activate the OLED and obtain an image point. The pixels are powered sequentially per row at the frame rate to obtain the image on the screen. The diodes (or pixels) are only lit for a short duration, corresponding to the image refresh time. Continuously powering the pixels causes resistive losses and substantial heat dissipation. In order to limit this loss of energy, which could degrade the diodes, the number of rows in the matrix is limited and the screen resolution is low.

PMOLED screens are relatively simple to produce and the production cost is low. On the other hand, the resolution of the images is poor. They are used for small screens for portable electronic equipment such as telephones or MP3 players.

Figure 1.16. *Switching circuit of a pixel of the screen: a) with PMOLED passive matrices; b) with AMOLED active matrices*

1.4.2.2. *AMOLED: active-matrix screens*

Unlike the PMOLED screen, the pixels or diodes of the AMOLED screen are lit in a controlled manner using the transistors associated with them. Associated with each pixel are two transistors, T_1 and T_2, functioning as switches, and a storage capacitance, C. By scanning the signal, the line is selected, closing the switch controlled by the transistor, T_1. The pixel's turn-on voltage, V_P, is transmitted to the base of the transistor, T_2. The switch controlled by the transistor, T_2, is in turn closed, allowing the pixel to be lit. The latter remains lit for the image refresh duration thanks to the voltage applied to the terminals of the capacitance, C. This mode of operation limits the power consumed by the screen and enables an increase in the number of pixels used for displaying the image. Consequently, this technology can be used for large and high-resolution screens, for example, high-frequency OLED television screens.

1.4.3. *OLEDs for automotive equipment*

The first OLED applications for automotive equipment were radio screens integrated in the dashboard. OLED screens currently equip several vehicle dashboard accessories, such as on-board computers or interior lighting. White OLEDs can be used for exterior (low-beam and high-beam) headlamps, which must meet precise specifications on flux, colorimetry and network geometric characteristics. OLEDs offer the advantage of flexible emitting films, allowing new headlamp designs to be developed, which would not be possible with the current automotive lighting sources. In addition, these lamps are compact and lightweight compared to other technologies and enable a reduction in the dimensions and volume of the optics. The problems associated with the use of OLEDs as automotive lighting sources are, first, the particularly severe conditions of headlamp use (humidity and mechanical vibrations), which may lead to premature degradation of the diode and consequently insufficient lifetime for the function to be fulfilled. Second, the production cost of OLED lamps remains high compared to those of other technologies, which does not satisfy the selection criteria for the automotive industry. These problems must be resolved so that OLEDs can become standard equipment in the industrial automotive field. Other projects aim to use transparent OLEDs to make a transparent car roof that lets in daylight. These same lamps are used to illuminate the car interior at night.

1.5. Conclusion

OLEDs have been studied since the 1990s and their applications were quickly marketed as single-screen portable electronic devices to replace LCD screens. Significant advances in material research, component design and understanding of physical processes have enabled the development and marketing of curved, thin and, more recently, rollable television screens. These screens are large and offer remarkable optical and electronic quality performance. In the field of lighting, decorative OLED lamps have also been marketed at affordable prices. The development of OLEDs for high-luminance lighting is progressing. In the automotive field, in spite of a new application perspective, production currently remains limited to screens for portable dashboard equipment.

We have not addressed the problems of degradation of OLEDs which constitute a technological barrier to be removed not only for LEDs but also for all organic devices. The physical processes of degradation of materials and organic components, and their consequences, are studied in Chapter 4.

2

Organic Solar Cells

2.1. Introduction

On the occasion of the COP (Conference of the Parties) 21, organized in Paris in 2015, a ranking – drawn up by survey – of the concerns felt by the world's population with regard to the coming 50 years placed the question of energy in first place. Naturally, energy has become an essential matter since the rapid development of industry in countries with near-infinite human potential and natural resources compared to those in developed countries. First, natural resources for fossil fuels (oil, coal and natural gas) are limited and risk being depleted in the not-too-distant future if the pace of consumption continues to rise steadily or indeed accelerates. Second, the threat of climate change that has already manifested itself in the form of violent weather phenomena in different regions of the world is becoming clearer with the massive increase in the concentration of carbon dioxide (CO_2) in the atmosphere. This *greenhouse gas* is produced by the combustion of raw materials during industrial and domestic activities. It is considered as a major factor in global warming due to its action of blocking the thermal dissipation of heat emitted from and by the Earth. Faced with these societal problems, and in order to protect the natural environment from the threat of contamination by nuclear waste, governments worldwide have encouraged efforts to develop renewable and non-polluting energy. Harnessing solar power seems to be the most efficient solution in order to meet this challenge. Not only is this energy source infinite because it is supplied by the Sun, but it is also available practically everywhere (with the exception of certain geographical areas); nor does harnessing this energy for use require the significant energy consumption that fossil-fuel sources do.

In this context, the development of solar cells for electricity generation is becoming a significant issue in ensuring energy autonomy for both industrialized and non-industrialized countries. At present, the material that has been most widely studied for solar cells is silicon. Thanks to its remarkable physical properties, this material has enabled the design of high-efficiency cells (~25%), yet its production cost is high and waste recycling remains an as yet unsolved problem. With the advent of organic semiconductors and further to the development of OLEDs, *organic solar cells* (OSCs) or organic photovoltaics (OPVs) have been studied to find a solution to harness solar energy without the disadvantage of costliness and possible pollution encountered with silicon-based cells.

In this chapter, we examine the physical processes of energy conversion by inorganic and organic solar cells. We look at the materials used as the OPV absorbent layer, their operation, their performance and their evolution within a market that is beginning to hold promise for the near future.

2.2. Solar spectrum

Solar radiation is emitted by thermal irradiation from the surface of the Sun. The solar spectrum is characterized by the *spectral energy flux density* $P(\lambda)$ (in units of $\mathrm{Wm^{-2}\,nm^{-1}}$), which is defined as the power of the radiation received per unit area and per unit wavelength. Moreover, we define the *photon flux density* $\Phi(\lambda)$ as the number of photons received per unit area, per unit wavelength and per second (in units of $\mathrm{m^{-2}\,nm^{-1}s^{-1}}$). The two characteristic quantities of the solar spectrum are linked by the relation:

$$\Phi(\lambda) = P(\lambda)\frac{\lambda}{h\,c} \qquad [2.1]$$

where h is Planck's constant and c is the speed of light in vacuum.

The Earth continuously receives solar radiation, with the amount of energy received varying depending on its position with respect to the Sun. The *solar constant* is defined as the total power received by a unit surface exposed perpendicular to the solar radiation and located at a distance of 1 au, that is, the average distance between the Sun and the Earth, that is, approximately 150×10^6 km.

This constant is equal to 1,360 Wm^{-2} and represents the energy flux at the surface of the Earth's atmosphere. However, on the Earth's surface, the power of solar radiation is attenuated due to absorption by the Earth's atmosphere. This power also depends on the position of the observer recording the spectrum, as well as on the local climatic conditions at the moment they take their measurements.

Energy from solar radiation on a region therefore depends on several parameters. To take these into account, we introduce the concept of *air mass* (AMx) in order to be able to consistently compare the results of measurements carried out under different geographical and climatic conditions. The air mass AM1 (x = 1) corresponds to the measurement conditions applied on the Earth's surface, with the sun at its zenith and in clear weather. When the sun is on an axis inclined at an angle θ with respect to the zenith direction and under the same climatic conditions, the air mass is:

$$\text{AM} = \frac{1}{\cos\theta} \qquad [2.2]$$

For the value $\theta = 60°$, the air mass is AM2. The air mass AM1.5 is frequently used in solar cell measurements and corresponds to the value $\theta = 48.2°$. For this air mass, the total power received on a unit surface (or the *irradiance*) placed on the ground is 1,000 W/m^2 or 100 mW/cm^2. The air mass zero, AM0, is also used to indicate the conditions under which the Sun is at its zenith and above the Earth's atmosphere.

Figure 2.1 illustrates the solar spectrum of a black body and that of the Sun, AM0 and AM1.5. It can be observed that the irradiance of the spectrum AM1.5 is high in the visible range and lower in the ultraviolet (UV) and infrared (IR) ranges.

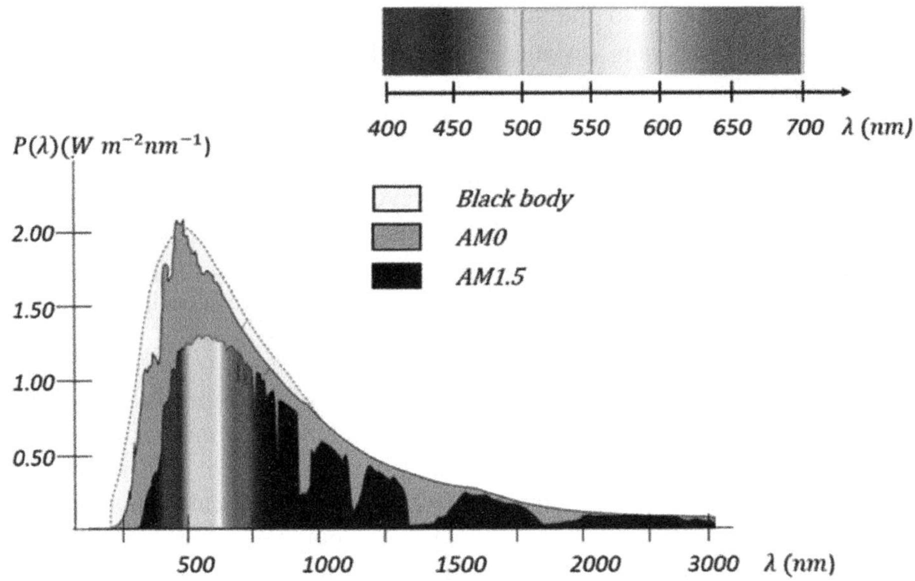

Figure 2.1. *Solar spectra for air mass AM0, AM1.5 and the emission spectrum of a black body; inset, the visible spectrum. For a color version of this figure, see www.iste.co.uk/nguyen/electronics2.zip*

2.3. Operating principle

When the light from the solar radiation reaches the Earth's surface, it is absorbed by the active material of the solar cells, which transform its energy into electrical energy through the *photovoltaic effect*. This effect is observed with the onset of a voltage at the terminals of devices exposed to light, regardless of their structures and the nature of the active material (organic or inorganic). It describes the energy conversion process in four stages:

– absorption of a photon by the active material and creation of an exciton;

– diffusion of the exciton created to the recombination site;

– dissociation of the exciton created into an electron and a hole;

– transport of charges created to the electrodes where they are collected.

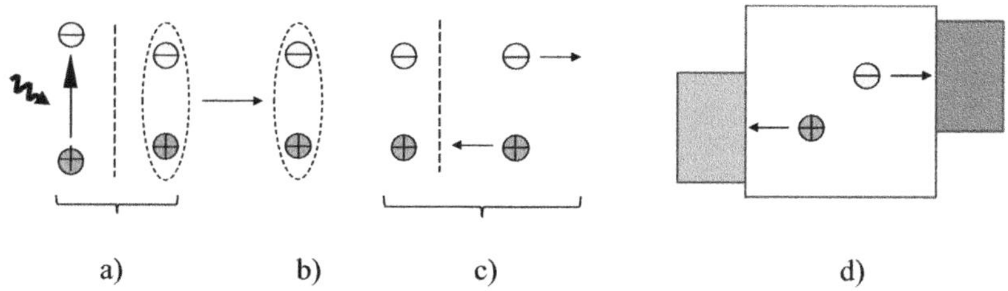

Figure 2.2. *The stages of the photovoltaic effect: a) absorption of a photon and creation of an exciton; b) diffusion of the exciton; c) dissociation of the exciton; d) transport and collection of charges at the electrodes. For a color version of this figure, see www.iste.co.uk/nguyen/electronics2.zip*

2.3.1. *Absorption of photons*

When the energy, $E = h\nu$, of the incident photon is higher than the gap, E_G, of the SC, it is absorbed by the latter and, by means of the *generation mechanism*, it enables the formation of an exciton.

For a direct bandgap SC, the *band-to-band transition* of the electrons is vertical and conserves the wave vector. The SC absorption coefficient in this case is:

$$\alpha(h\nu) = A \times (h\nu - E_G)^{1/2} \quad [2.3]$$

For an indirect bandgap SC, the band-to-band transition of the electrons is oblique, and the wave vector is not conserved, as part of the energy, E_P, is lost through thermalization with the emission of a phonon. The SC absorption coefficient in this case is:

$$\alpha(h\nu) = A \times \frac{(h\nu - E_G - E_P)^2}{1 - e^{-E_P/kT}} \quad [2.4]$$

The absorption coefficient depends on the SC gap and is all the greater the smaller the gap is. It is therefore advantageous to choose a narrow-gap (or low bandgap) SC as the active material of the solar cell. Silicon has a gap of $E_G = 1.12$ eV and constitutes an excellent absorber for P–N junction cells. Moreover, the intensity of the solar spectrum is significant in the range of wavelengths between 600 and 1,000 nm (corresponding to the energy range of 1.3–2.0 eV). Therefore, it is of interest to use absorbers with a gap, E_G, having a value within this range to optimize or improve photon absorption. For solar cells based on conjugated polymers, the gap for common materials is greater than 2.0 eV, which explains the low efficiency of these cells.

On the other hand, the intensity of light absorbed by the material decreases exponentially with the depth of penetration (see equation [3.4] in Volume 1). Knowing the absorption coefficient $\alpha(h\nu)$, the absorption depth can be estimated for each wavelength of the solar spectrum. Ultraviolet radiation is absorbed over a small film thickness, while infrared radiations penetrate the material more deeply prior to full absorption of the light. For silicon, a thickness of a few hundred micrometers is required to absorb solar radiation, whereas for organic materials, absorption is virtually complete for films approximately 100 nm thick. In some cases, the use of very thin films is recommended for organic cells to compensate for the low mobility of carriers. However, using such films may result in the loss of photons that pass through the material without being absorbed and without creating excitons and therefore decrease the cell efficiency.

2.3.2. *Diffusion of excitons*

Once formed on a site in the active material, the exciton moves to neighboring sites by means of an energy transfer mechanism or by jumps. The diffusion distance is of the order of a few tens of nanometers for excitons created in an organic SC and can reach several tens of micrometers in an inorganic SC. This is an average distance because during their diffusion, exciton charges can be captured by traps or recombine before reaching sites that enable their separation. The disappearance of excitons, constituting a loss of efficiency, is due to the presence of trap sites or the morphology of the material. To minimize these losses, purified materials should be used, that is, with few or no defects and which are structurally ordered, enabling excitons to diffuse over a long distance, greater than that separating two sites. On average, for organic SCs, the diffusion length of carriers in small molecules and conjugated polymers is of the order of 10 nm (Peumans *et al.* 2003). The size of the donor/acceptor domains of the absorber material should then be of the same order of magnitude to optimize the diffusion process, which allows effective separation of charges.

2.3.3. *Dissociation of excitons*

The charges of an exciton are bound together by the Coulombic attraction force, which is likely to be broken by an energy supply greater than their binding energy. In this case, charges are separated and can move through the material to the appropriate sites. As the binding energy of excitons is significant, especially in organic SCs, thermal agitation in the material is not sufficient to dissociate excitons.

In the case of a P–N junction solar cell, the dissociation of excitons is induced by the internal electric field in the SCR. The charges are separated by the field and drained into the SCs: holes in the field direction, that is, towards the P-type SC, and electrons in the opposite direction, that is, towards the N-type SC.

In an organic SC, the concept of the P–N junction is not suitable owing to the dimensions of the materials and devices. Indeed, the thickness of organic films is of the order of a few tens of nanometers and the diffusion length of the carriers is approximately 10 nm on average. In order to achieve efficient carrier dissociation, an *electron donor material (D)* and an *electron acceptor material (A)* are used to form the absorber material. At the nanoscale, the association of *(D/A)* materials constitutes a P–N junction and creates an electric field enabling the dissociation of the exciton charges. When an exciton reaches the interface between a donor and an acceptor, the potential difference between the LUMO levels induces an electric field that can dissociate the exciton into an electron that becomes free in the LUMO band of the acceptor, and a free hole in the HOMO band of the donor.

The process of exciton dissociation is crucial for the photovoltaic effect because it enables the electrons to be separated from the holes that will constitute charge reservoirs for the solar cell. The first organic cell structures used the concept of the P–N junction of conventional solar cells (Tang 1986). These cells, called *bilayer cells*, are formed from the assembly of a donor layer and an acceptor layer in contact over a surface, *S*. The separation of charges in these devices takes place at the interface of the layers, that is, across the entire contact surface area. As a result, excitons created at any site in the active material must diffuse to the interface to be dissociated and they are unlikely to reach this point as their lifetime is short and the diffusion length is usually approximately 10 nm. Only excitons created in the immediate vicinity of the interface will be able to access it and the number of dissociated excitons is small. In this case, few charges are collected on the electrodes.

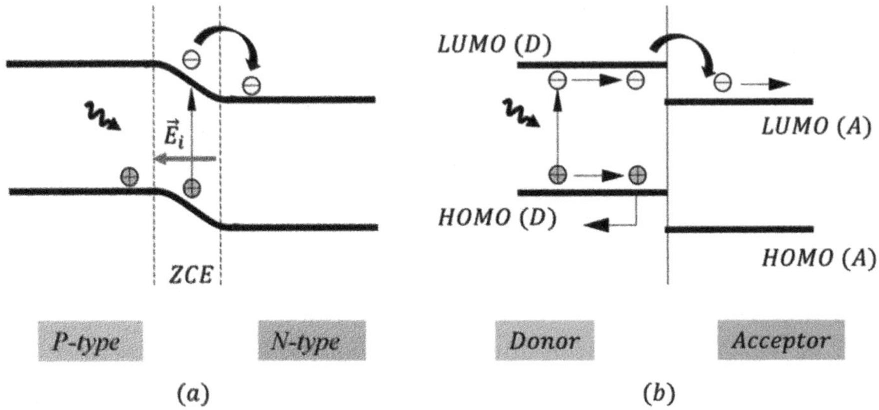

Figure 2.3. *Dissociation of charges in a) a P–N junction solar cell; b) an organic solar cell. For a color version of this figure, see www.iste.co.uk/nguyen/electronics2.zip*

In order to increase the exciton dissociation rate in the absorber, a new concept has been proposed, with the cell structure known as the *structure of interpenetrating networks* or *bulk heterojunction* (BHJ). The absorber material is a nanocomposite obtained by mixing a donor material and an acceptor material. The size of the D and A domains is of the same order of magnitude as the diffusion length of the carriers to maximize the probability of exciton dissociation at the donor/acceptor interface. Furthermore, the concentrations of donors and acceptors must be sufficient to ensure good continuity of the domains between electrodes, which contributes to improved transport and, consequently, collection of carriers. BHJ structures present an obvious advantage over bilayer structures in that the contact surface between the D and A domains is considerably higher due to the nanostructure of the absorber. The P3HT:PCBM composite is a widely studied, efficient absorber for BHJ structures. The conversion efficiency of cells using this composite reaches 6.5%, and certain composites of derivatives of P3HT and PCBM are used as absorbers in solar cells with an efficiency close to 9% (Lu *et al.* 2015).

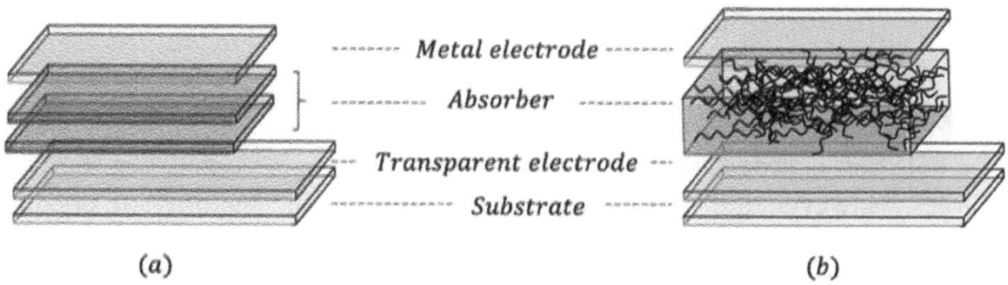

Figure 2.4. *Concepts of solar cell structure: a) bilayer structure; b) bulk heterojunction structure. For a color version of this figure, see www.iste.co.uk/nguyen/electronics2.zip*

The main benefit of the BHJ structure lies in mixing the *D* and *A* domains at the nanoscale, which enables an increase in the contact surface area, and therefore in the rate of dissociation of excitons created. However, because of their small size, the domains tend to aggregate and the composite dissociates into different *D* and *A* zones, decreasing the contact surface area and the rate of exciton dissociation. In this case, a *phase separation occurs* in the BHJ, and the consequence of which is a decrease in the number of dissociations in the amount of charges collected at the electrodes. To avoid this disadvantage, one component material (often the acceptor) is used in the form of an organized network of nanorods or nanorod arrays and the other material (donor polymer) is deposited in solution, filling the space between the rods. The composite formed is uniform and stable and the contact surface area between the *D* and *A* materials can be controlled and modified by controlling the nanorod network.

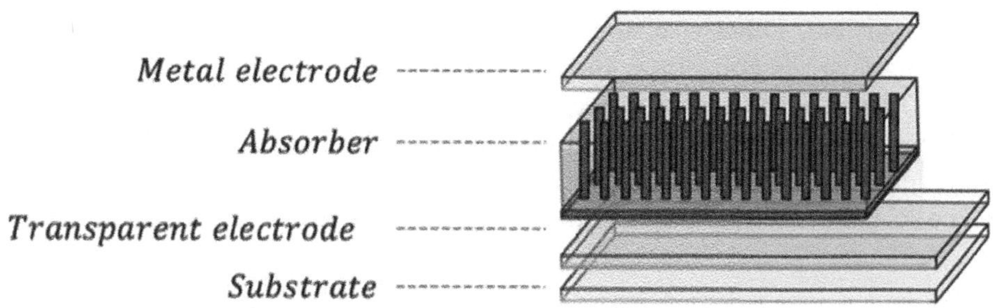

Figure 2.5. *Structure of a solar cell using an absorber comprising nanorod arrays. For a color version of this figure, see www.iste.co.uk/nguyen/electronics2.zip*

2.3.4. *Diffusion of carriers to electrodes*

After dissociation of the exciton, free charges can either recombine or move towards the electrodes. Recombination is possible between a free electron and a free hole and is produced by the Coulombic attraction force between the two charges. However, the recombination time range in the majority of organic materials (of the order of a microsecond) is lower than that of their dissociation (of the order of a picosecond) so that free carriers can reach the electrodes before they recombine. They move towards the electrodes under the influence of the electric field created in the material by the potential difference between electrodes with different work functions. Electrons move towards the low work-function electrode (cathode) and the holes towards the high work-function electrode (anode). The movement of the charge carriers is also facilitated by the difference in concentration between the various domains of the material. High-concentration electrons in the LUMO band of the acceptor-material part located near the donor/acceptor interface diffuse to the region of the same material with a lower concentration. Similarly, high-concentration holes in the donor HOMO band at the donor/acceptor interface diffuse to the low-concentration region of the donor material. To minimize losses due to transport, carriers' mobility should be high enough to be able to reach the electrodes before their end of life and be effectively collected there.

2.3.5. *Collection of charges*

The charges are collected at the SC/electrode interface. The carriers pass from the SC to the electrode across the potential barrier at the interface. Ideally, the SC/electrode contact needs to be ohmic to facilitate the passage of the carriers. Otherwise, the potential barrier should be low in order to maximize carrier collection. In a simple organic solar cell structure, if the LUMO level of the acceptor is close to the Fermi level of the cathode metal and the HOMO level of the donor is close to the Fermi level of the anode conducting oxide, the contacts are considered as ohmic and the carriers can be collected at the electrodes with no or minimal loss.

2.3.6. *Process optimization for an organic solar cell*

The overall photovoltaic effect processes involved in an organic solar cell are summarized in Figure 2.6. In this diagram, the cell shown is a complete structure comprising electron and hole transport layers.

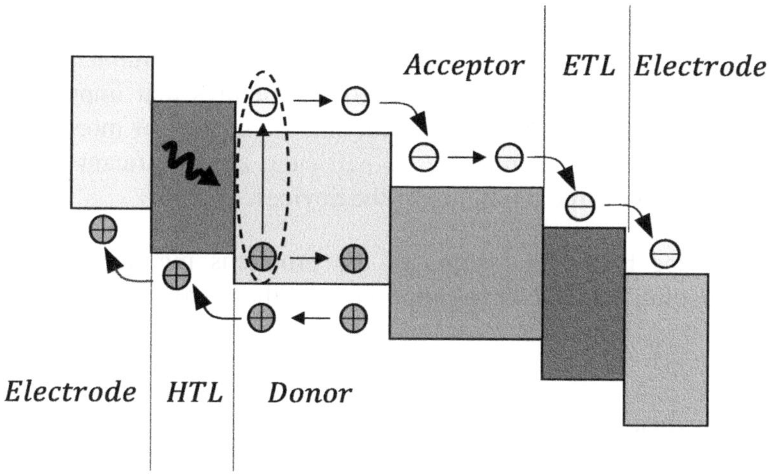

Figure 2.6. *Photovoltaic effect in a complete solar cell. For a color version of this figure, see www.iste.co.uk/nguyen/electronics2.zip*

Ideally, each photon absorbed by the cell should generate an exciton whose dissociation will provide an electron and a hole, which will be collected at the electrodes. However, we have seen that the optical and electronic processes that take place in the different parts of the cell do not enable the photovoltaic effect to be fully realized. Solutions do nevertheless exist to optimize the operating conditions of the cell in order to obtain good conversion of light energy into electrical energy. The cell's performance can be improved by considering the following conditions:

– the range of wavelengths absorbed by the absorber must be as wide as possible with respect to the solar spectrum;

– the recombination losses should be minimized;

– the exciton dissociation should be maximized;

– the number of charges transported to the electrodes should be maximized.

At present, the performance of organic solar cells remains modest because the conditions listed have not been met. The sunlight wavelengths absorbed by the cells are still limited by the optical characteristics of organic materials. The quality of the materials used does not yet enable minimization of the defects that are at the origin of the centers of recombination or traps that affect carrier transport. Similarly, BHJ structures that improve exciton dissociation have not yet been mastered, and the control of morphology and stability of donor/acceptor mixtures in particular are significant problems to be resolved to ensure the reliability of the devices.

We will see below that some of the problems mentioned have been studied and solutions have been proposed, and these have offered interesting results.

2.4. Characteristic parameters of solar cells

2.4.1. *Current–voltage characteristics*

In the dark, a solar cell behaves like a diode and the current density–voltage characteristic is that of a diode. The current density equation for a P–N junction diode subject to voltage V is called the *Shockley equation* and is written as:

$$J_{ob}(V) = J_S \left[exp\left(\frac{|e|\,V}{k\,T}\right) - 1\right], \qquad [2.5]$$

where J_S is the saturation current density.

Under illumination, a photocurrent is created and is added to the previous current, which becomes:

$$J(V) = J_S \left[exp\left(\frac{|e|\,V}{k\,T}\right) - 1\right] - J_{ph} \qquad [2.6]$$

Equation [2.6] represents the characteristic of an ideal diode, not taking account of residual resistances. For an actual diode, the losses are represented by two resistances: 1) the series resistance, R_S, which represents the contact resistance and depends on the resistivity of the material, that of the electrode and of the Schottky barrier upon material/electrode contact; and 2) the parallel resistance, R_P , which represents the leakage current in the diode.

The equivalent circuit of the solar cell under illumination is given in Figure 2.7 and the equation of the $J(V)$ characteristic becomes:

$$J(V) = J_S \left[exp\left(\frac{|e|\,(V - J\,S\,R_S)}{\eta\,k\,T}\right) - 1\right] + \frac{V - J\,S\,R_S}{R_P} - J_{ph}\,, \qquad [2.7]$$

in which S is the cross-section of the resistance R_S and η is the ideality factor of the diode.

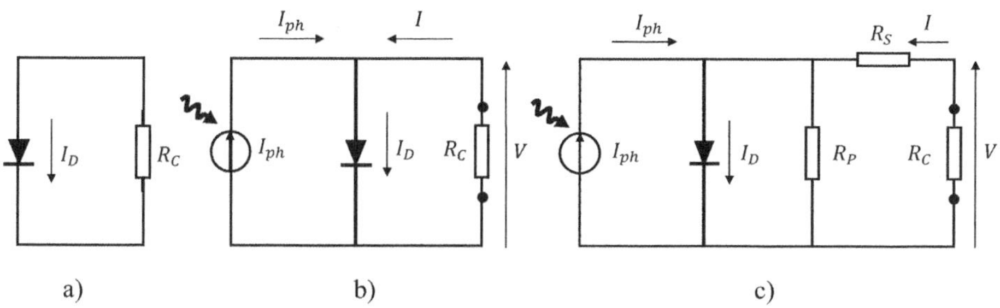

Figure 2.7. *Equivalent circuit of a solar cell: a) ideal cell in the dark; b) ideal cell under illumination; c) actual cell under illumination*

Experimentally, the current–voltage characteristics in illuminated organic solar cells do not necessarily follow expression [2.7], and the photocurrent, J_{ph}, may depend on the electric field in the cell (Deibel and Dyakonov 2010). This dependence can be explained by the processes of formation and dissociation of excitons and extraction and transport of free charges described in section 2.3. These processes depend on the apparent electric field in the different layers of the cell. In some cells, the maximum value of J_{ph} is greater than the short-circuit current density, J_{SC}, which invalidates expression [2.7], established for inorganic cells. Therefore, it is necessary to take account of the dependence of J_{ph} on the electric field, which can play more or less of an important role depending on the nature of the active material and the structure used for the cell. With this specified, the equivalent circuit in Figure 2.7 can be applied to describe the general operation of organic solar cells.

2.4.2. *Photovoltaic parameters of a solar cell*

The photovoltaic parameters of a solar cell are determined from the current–voltage characteristics of the illuminated cell. The $J(V)$ characteristic of the illuminated cell lies away from that measured in the dark and enables the photovoltaic parameters of the cell to be determined.

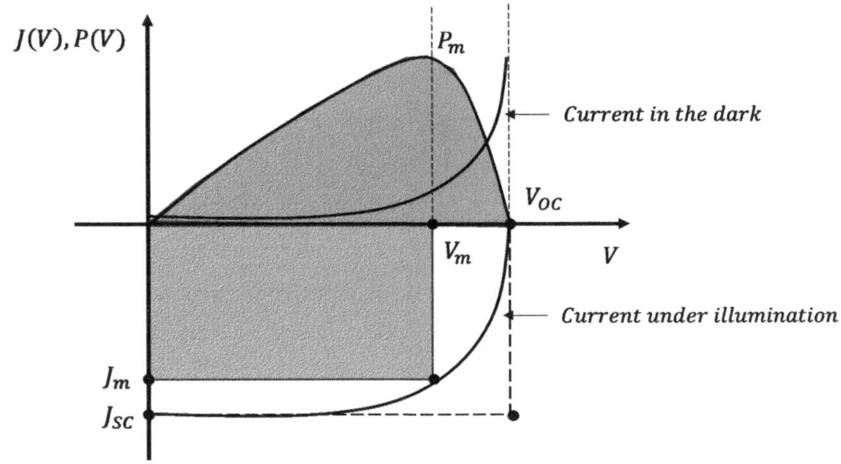

Figure 2.8. *$J(V)$ characteristics and photovoltaic parameters of a solar cell in the dark and under illumination*

2.4.2.1. *Short-circuit current (I_{SC})*

The short-circuit current I_{SC} is the current measured in the external circuit when the cell electrodes are short-circuited, thus at zero voltage. For an ideal cell, the short-circuit current density is equal to the photocurrent density according to the relation [2.6]:

$$J_{SC} = -J_{ph}$$

This equality is observed for many organic cells, but in cells where the electric field notably influences the physical processes of excitons and free charges, we have:

$$J_{SC} > -J_{ph}$$

The short-circuit current depends on the density of the incident luminous flux and the absorption of solar radiation by the active material, that is, the gap of the organic SC. To increase the short-circuit current, one possible solution is to use narrow-gap absorbers to extend the absorption spectrum up to, and beyond, the red and near-infrared solar radiation range to make best use of the solar spectrum and to collect maximum luminous flux from the Sun. According to expression [2.6], in order to increase the short-circuit current, the shunt resistance must be large to minimize the leakage current and the series resistance must be low to minimize losses.

2.4.2.2. *Open-circuit voltage*

The open-circuit voltage, V_{OC}, is the voltage measured at the cell terminals when the current is zero in the external circuit. This voltage is the maximum voltage delivered by the cell and corresponds to the voltage at which the current in the dark counterbalances the current under illumination. It is expressed, as a function of the saturation current and the photocurrent, by the expression:

$$V_{OC} = \frac{k\,T}{|e|} ln\left(\frac{J_{ph}}{J_S} + 1\right) \quad [2.8]$$

Since variations in the photocurrent are generally small, the open-circuit voltage, V_{OC}, depends essentially on the saturation current, which in turn depends on the recombination process. In other words, the open-circuit voltage enables assessment of the importance of recombination in the solar cell studied. In a solar cell with a BHJ structure, the open-circuit voltage depends on the energy difference between the donor's HOMO level and the acceptor's LUMO level, and the choice of the donor/acceptor pair would enable its value to be optimized. The difference between these two energy levels is called the *effective gap* $E_{G,DA}$ of the donor/acceptor system of the BHJ structure:

$$E_{G,DA} = E_{HOMO,D} - E_{LUMO,A} \quad [2.9]$$

An empirical relationship was proposed for the expression of the open-circuit voltage, V_{OC}, following an experimental study of BHJ cells using PCBM as the acceptor material (Scharber *et al.* 2006):

$$V_{OC} = \frac{1}{|e|}\left(E_{HOMO,D} - E_{LUMO,PCBM}\right) - 0.30 \quad [2.10]$$

To improve the open-circuit voltage, V_{OC}, the choice of donor and acceptor materials should be made in such a way as to increase the effective gap while ensuring that narrow-gap materials are promoted to increase the absorption of light radiation. For example, the V_{OC} value of cells using the PCBM mixture is typically 0.50–0.60 V with P3HT and 0.80–0.90 V with poly(2-methoxy-5-(3',7'-dimethyl-octyloxy)-1,4-phenylene-vinylene) (MDMO-PPV) (Sariciftci 2004). Other factors also influence the open-circuit voltage, V_{OC}, and, in particular, the interfaces at the electrodes that are responsible for charge losses and, therefore, decrease V_{OC}. Using the ETL and HTL transport layers can enable an increase in the open-circuit voltage, V_{OC}.

S-shaped current–voltage characteristics and low conversion efficiency are observed in some cells. These particular characteristics would be due to the transport mechanism limited by the space charge accumulated at the electrode/absorber interfaces (Wagenpfahl *et al.* 2010).

2.4.2.3. *Fill factor*

The J_{SC} and V_{OC} parameters mark the boundary values of the current density and voltage but do not enable assessment of the maximum power delivered by the cell studied. This power, denoted as P_m, is obtained when the product of the voltage V_m by the current density J_m is maximum, the values V_m and J_m being lower than the boundary values, V_{OC} and J_{SC}, respectively, due to the losses (see Figure 2.8). Some studies propose general relationships linking the V_m, J_m, V_{OC} and J_{SC} parameters based on experimental results (Nayak *et al.* 2019). Depending on the structure of the active layer (bilayer or BHJ), the relationships are different; and for the BHJ structure, it is accepted that the maximum value of the open-circuit voltage, V_{OC}, is determined by the effective gap of the donor/acceptor mixture.

The *fill factor* (FF) is defined by the following expression:

$$FF = \frac{J_m \times V_m}{J_{SC} \times V_{OC}} \quad [2.11]$$

This parameter compares the characteristic voltage and current density values with those corresponding to the maximum power produced by the cell. It gives a description of the quality of the *J(V)* curve under illumination. As $J_m < J_{SC}$ and $V_m < V_{OC}$, we have $FF < 1$, and it is important to have a fill factor close to unity to obtain high cell power. If the photocurrent increases sharply as the voltage approaches the open-circuit voltage, V_{OC}, according to Figure 2.8, the fill factor will be high. From the $J(V)$ characteristic, the J_{SC} and V_{OC} parameters can be determined directly but the $P(V) = J(V) \times V$ curve must be plotted to determine the values of J_m and V_m in order to obtain the fill factor. Assuming that the solar cell is an ideal diode and calculating the derivative of the power delivered by the cell $P(V) = J(V) \times V$ with respect to the voltage V, the voltage V_m can be determined. An approximate calculation enables the fill factor to be expressed as a function of the voltage V_{OC} (Bisquert 2018):

$$FF = \frac{\frac{|e|\, V_{OC}}{\eta\, k\, T} - ln\left(1+\frac{|e|\, V_{OC}}{\eta\, k\, T}\right)}{1+\frac{|e|\, V_{OC}}{\eta\, k\, T}} \qquad [2.12]$$

In this model, the FF factor depends only on the ideality factor, η, and the open-circuit voltage, V_{OC}.

The J_{SC}, V_{OC} and FF parameters of an organic solar cell can be improved by optimizing the film deposition method. Controlling manufacturing conditions to optimize the morphology or thermal treatment of films and cells can significantly increase efficiency. However, the values of the short-circuit current and open-circuit voltage are limited by the properties of the absorber material. Therefore, obtaining a good-quality organic material is necessary to produce a solar cell with high conversion efficiency.

2.4.3. *Efficiency*

External quantum efficiency (EQE), or incident photon to current efficiency (IPCE), is defined as the ratio of the number of electrons in the external circuit to the number of incident photons of wavelength λ on the cell surface:

$$\eta_{ex} = \frac{number\ of\ photo\text{-}generated\ and\ collected\ electrons}{number\ of\ incident\ photons} \qquad [2.13]$$

For wavelength λ, we can write the expression of the short-circuit current density as:

$$J_{SC}(\lambda) = \eta_{ex} \times \frac{|e|\, \lambda}{h\, c} \times P_{in}(\lambda), \qquad [2.14]$$

where P_{in} is the luminous intensity per unit area illuminated.

In the range of wavelengths absorbed by the material, limited by the λ_{min} and λ_{max} values, we have:

$$J_{SC}(\lambda) = \frac{|e|}{h\,c}\int_{\lambda_{min}}^{\lambda_{max}} \eta_{ex} \times P_{in}(\lambda) \times \lambda\, d\lambda \qquad [2.15]$$

The value of the current density, J_{SC}, can be estimated with the knowledge of the efficiency, η_{ex}, of the solar cell and the spectrum of incident radiation.

For all solar cells, the performance parameter is that which measures the power supplied by the device. This parameter is the *power conversion efficiency* (PCE), η_P, defined as the ratio of the power P_{max} supplied by the cell to the incident luminous power, P_{in}:

$$\eta_P = \frac{P_{max}}{P_{in}} = FF \times \frac{V_{OC} \times I_{SC}}{P_{in}} \qquad [2.16]$$

This efficiency is a function of the fill factor, the open-circuit voltage and the short-circuit current. Generally, the value of the efficiency, η_P, of a solar cell is determined for the standard air mass 1.5 AM spectrum in which the power, P_{in}, has a value of 1 kW m^{-2} or 100 mW cm^{-2}.

In a P–N junction cell, the maximum conversion efficiency theoretically evaluated by Shockley and Queisser (1961) reaches approximately 30% for an SC with an energy gap of 1.10 eV (silicon).

For a bulk heterojunction organic cell, the efficiency limit is approximately 23% (Kirchartz *et al.* 2009). However, the best results reported for BHJ cells are much lower (~ 12%) (Liu *et al.* 2015) than the theoretical value.

This difference is explained by the various losses in the photovoltaic effect stages, which are as follows:

– optical losses due to the absorption of photons by the non-active layers of the cell;

– losses of excitons that do not reach the donor/acceptor interface;

– losses of charge carriers due to non-radiative recombination at the interfaces of the layers and by the traps in the absorber;

– losses of charge carriers that do not reach the electrodes to be collected.

To further improve the performance of organic solar cells, it will be necessary to limit the losses listed by optimizing the parameters concerning the active layer, transport layers and electrodes, thereby enabling the cell architecture to be structured effectively. Future development of these cells will focus on the design and synthesis of new, high-quality organic materials, which will enable loss problems to be resolved and cell efficiency to be improved.

2.5. Organic materials

The materials studied in this section are those used as electron donors or acceptors in the active layer in organic solar cells.

2.5.1. *Electron donor materials*

Electron donor materials (*D*) have a low ionization potential and are likely to lose an electron in contact with an acceptor material (*A*) that has a high electron affinity and therefore is likely to gain an electron. Donor and acceptor materials can be small molecules or polymers. The advantage of using small molecules lies in the possibility of controlling the structure and the molecular weight of the material, which enables the production of layers with more reliable and better controlled physical characteristics than polymers.

An efficient donor material must meet a number of criteria. First, for the efficient dissociation of excitons, the difference between the HOMO levels of the donor and the acceptor must be greater than 0.3 eV (Scharber *et al.* 2006). Furthermore, as the open voltage, V_{OC}, depends on the effective gap according to relation [2.9], the donor HOMO level needs to be as low as possible with respect to the acceptor LUMO level. In the active layer of the cell, the donor material also ensures the transport of holes and their mobility must be sufficient to allow efficient charge collection at the anode. Finally, the material must form as homogeneous a blend as possible with the acceptor, in order to optimize the charge dissociation at the D/A interfaces.

Let us give a brief description of the most widely used donor materials in solar cells:

– phthalocyanine (Pc) donors: among these materials, copper phthalocyanine (CuPc) complex is frequently used in cells in association with an acceptor such as the fullerene C_{60}. The measured efficiency of these cells is ~3.5% (Uchida *et al.* 2004). Zinc complex (ZnPc) is also chosen as a donor material for its stability and the high exciton diffusion length;

– polyacene donors: the most widely studied polyacenes in organic electronics are pentacene and anthracene. These small molecules have a particularly high mobility (>0.1 $cm^2V^{-1}s^{-1}$) compared to the majority of organic materials, and are of interest for use as the active layer in organic transistors. For example, pentacene- and fullerene-based solar cells, C_{60}, have an efficiency of approximately 2.7% (Yoo *et al.* 2004). It should be noted that donor–acceptor mixtures can be deposited by thermal evaporation or in solution (often using polyacene derivatives);

– oligothiophene donors: these materials have high mobility and optical properties that can be adjusted by modifying the energy levels. Alpha-sexithiophene (6T) is an oligothiophene that is widely used for the active layer of organic electronic devices. Combined with the fullerene C_{70}, the mixture 6T :C_{70}, used as an absorber, enables an efficiency of ~ 2.38% to be obtained (Sakai *et al.* 2009). Other soluble oligothiophene derivatives can be deposited in solution and are used with the fullerene C_{61} in solar cells with an efficiency reaching 6.1% (Li *et al.* 2012);

– donors based on triphenylamine derivatives (TPA): these materials have high conductivity and absorption with low ionization potential. A donor material (DTDCTP) of donor–acceptor–acceptor (D–A–A) configuration using derivatives of TPA is associated with the fullerene C_{70} to form the absorber material of solar cells operating with an efficiency of 6.4% (Chiu *et al.* 2012). The alternating D–A configuration, called *push–pull,* makes it possible to obtain a narrow energy gap, which improves absorption of the material and will be explained below;

– polymer donors: P-type conjugated polymers, such as PPV and its derivatives, have been widely used as donor material. However, due to their energy gap, photon absorption is limited and cell efficiency remains poor (< 3%). With its relatively narrow gap (~ 2 eV) and good hole mobility $\left(\sim 0.1\ cm^2V^{-1}s^{-1}\right)$, enabling greater performance to be obtained, P3HT has become the most widely studied conjugated polymer for BHJ structures.

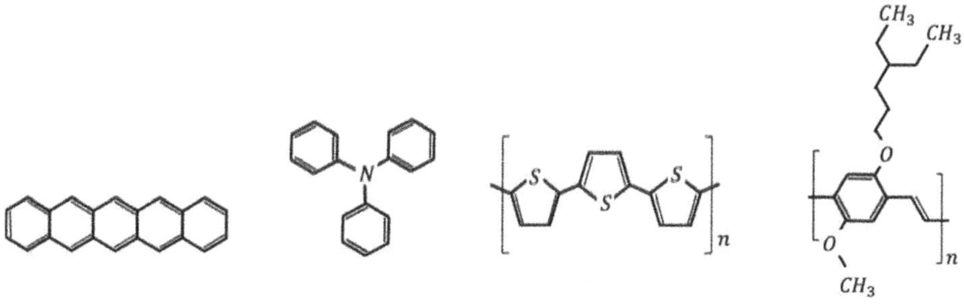

Figure 2.9. *Examples of electron-donor organic materials: a) pentacene; b) triphenylamine (TPA); c) oligothiophene; d) poly[2-methoxy-5-(2-ethyl-hexyloxy)-1,4-phenylene-vinylene] (MEH-PPV)*

2.5.2. *Electron acceptor materials*

Two types of electron acceptor materials exist: fullerenes and non-fullerenes.

2.5.2.1. *Fullerene-based acceptors*

The fullerene C_{60} and its derivatives constitute the most widely used acceptor materials for BHJ cells. Their main advantages are, first, a high conductivity, which favors the transport of charges, and, second, their ability to receive electrons from the donor material, which favors the transfer of charges in the exciton dissociation process. The simplest fullerenes are C_{60} and C_{70}. They can be deposited by evaporation under vacuum.

The derivatives of these fullerenes are $PC_{61}BM$ and $PC_{71}BM$, which are soluble and can be mixed with donor polymers and deposited in thin films by spin coating. $PC_{61}BM$ or [6,6]-phenyl-C_{61}-butanoate consists of a unit of fullerene C_{60} carrying a phenyl group and a chain of methyl butyrate, which enable its solubility in the usual solvents to be increased. The mobility of the charge carriers is estimated at approximately 0.2 $cm^2V^{-1}s^{-1}$, which is very high for an organic material. $PC_{71}BM$ is the derivative of the fullerene C_{70} and has properties similar to those of $PC_{61}BM$, but its visible absorbance is better due to its dissymmetrical structure.

2.5.2.2. *Non-fullerene acceptors*

We can distinguish two types of materials:

– non-fullerene acceptor materials, acceptors based on perylene diimide (PDI) or perylene-3,4,9,10-tetracarboxyl diimides and its derivatives: these materials are N-type SCs, which have suitable properties for the acceptor function such as high absorption of the visible spectrum, high electron mobility and good chemical and thermal stability. The solubility of these materials can be facilitated using alkyl side chains introduced on the imide nitrogen atoms. To enhance the electron accepting properties of the material, electron withdrawing groups can be added to the periphery of the aromatic parts. Materials similar to PDI, naphthalene diimide (NDI) and its derivatives, are used as the acceptor material but the cell efficiency is not as high because photon absorption is low due to the bandgap, which is wider than that of PDIs;

– N-type conjugated polymers: these materials have wide molecular engineering capabilities that allow them to be adapted to the energy levels in donor/acceptor mixtures. As an example, poly(cyano-phenylene-vinylene) (CN-PPV), containing a nitrile group, has been used as an acceptor material in combination with MEH-PPV, used as a donor to produce a cell entirely composed of polymers (Yu and Heeger 1995). However, the efficiency of solar cells using these materials remains low (0.9%), although the principle of mixing two polymers is of interest for obtaining a broadened absorption spectrum of the blend material.

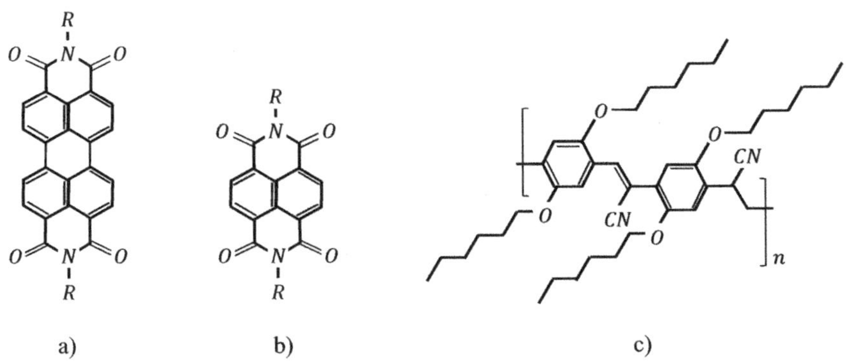

Figure 2.10. *Examples of non-fullerene acceptor materials: a) perylene diimide (PDI); b) napthalene diimide (NDI); c) CN-PPV*

2.6. P3HT:PCBM

Of the donor/acceptor systems used as the active material of organic solar cells, the P3HT:PCBM mixture is the most widely studied owing to its various aspects and the problems posed, which constitute barriers in the study of BHJ structures. The energy conversion efficiency of P3HT: PCBM-based cells is greater than 5%, that is, less than that of cells using certain organic materials as an absorber. However, the material has generated much interest because the charge transfer mechanism was first discovered in conjugated polymer/fullerene composites (Sariciftci *et al.* 1992), highlighting this particular material and its potential for application in organic electronics. Indeed, C_{60} has several intrinsic advantages. First, its LUMO level is relatively low with respect to the vacuum level and enables it to accept electrons from a P-type material. Second, the transfer time of an electron from the conjugated polymer to the fullerene is very short (~ 45 fs), which promotes charge separation. Finally, electron mobility is high, enabling efficient transport of the carriers to electrodes. The P3HT associated with the fullerene was studied with a view to replacing PPV and its derivatives in the early 2000s (Schilinsky *et al.* 2002), because it presents many advantages. First, the polymer absorption spectrum limit lies in the vicinity of 650 nm, which is close to the maximum of the solar spectrum, while that of PPV derivatives is centered at 550 nm. Second, the charge transport is high thanks to the high mobility of the holes (~ 0.1 $cm^2V^{-1}s^{-1}$). Both materials are soluble in the same organic solvents and can be deposited in homogeneous films for the solar cell active layer. The P3HT:PCBM mixture in principle possesses all the advantages to form an ideal donor/acceptor system for BHJ structures.

However, extensive studies of P3HT:PCBM-based cells have revealed numerous problems related to both the materials and the structure of the mixture. For the fullerene, solubility in common solvents is poor and it is necessary to use specific solvents. Furthermore, absorption in the visible is average because of the symmetrical structure of the molecule, which prohibits low-energy transitions. For P3HT, film deposition presents two problems. First, unlike the fullerene, which has a well-defined structure and consistent quality, regardless of the supplier, the quality of P3HT is variable, according to the synthesis method used. Using the material supplied by different manufacturers, for the same announced purity, and applying the same method of preparation (solvent, concentration, polymer quality and deposition technique), the physical characteristics of the deposited films are generally found to be different. Furthermore, using the same polymer supplied by the same manufacturer, the characteristics of the films obtained are different for different batches of products. It is therefore difficult to compare P3HT films created by different groups, given the homogeneity issues of the original polymer. The second problem with P3HT also lies in the solubility of the polymer, which can be used with several common organic solvents. However, the characteristics of the films are modified with the solvent used and the solvent suitable for dissolving both the polymer and the fullerene should be carefully chosen to prepare the P3HT:PCBM film. Usual and suitable solvents for the mixture preparation are chloroform, chlorobenzene and di/tri-chlorobenzene. Using these solvents, the polymer and the fullerene are simultaneously dissolved and form a uniform mixture in solution. Analyses show that ordered mixing domains are formed in the entire volume of the deposited film and the polymer chains, which are well organized, contribute to better transport of charges in the mixture (Reisdorffer *et al.* 2012). The use of an inappropriate solvent leads to inhomogeneous dissolution of the materials, and the fullerene, generally poorly soluble, tends to form clusters after the formation of the film. These problems relating to the structure of materials and their solubility lead to mixtures with characteristics that are difficult to control in practice, and therefore difficult to compare.

Regarding the optimum concentration of materials to obtain the best conversion efficiency, several P3HT:PCBM concentration ratios have been found and proposed. These can vary from 1:0.8 to 1:1 depending on the materials used, and there is no overall consensus on these ratios.

Annealing of active films is an important factor in improving cell efficiency. For the polymer, the annealing temperature varies between 100 and 150 °C and has the effect of rearranging the chains, which become more ordered (Chirvase *et al.* 2004). For the fullerene, annealing promotes the diffusion of molecules that tend to group and form aggregates in the mixture (Erb *et al.* 2005). A *solvent vapor annealing* (SVA) technique exists that allows the film to be kept at room temperature during annealing. It involves controlling the growth of the film by controlling the time needed for the solvent to fully evaporate and the film to dry. For a long annealing time, this technique enables self-organization of the molecules to obtain the optimum structure enabling high efficiency (Li *et al.* 2005a). Overall, annealing changes the morphology of the film, promotes the arrangement of the polymer chains, and increases the crystallinity and the transport of charges, which, in short, improves cell efficiency (Verploegen *et al.* 2012).

Despite all of the parameter optimizations to improve film structure and morphology, the efficiency of P3HT:PCBM-based cells appears to be reaching its limit, and other concepts and materials have been proposed to replace this material. Among the promising new materials for solar energy conversion, perovskite occupies a privileged place for the future development of flexible cells.

2.7. Perovskite

Halogenated perovskite used as an absorber in solar cells is a hybrid material with the general formula, ABX3, where *A* is an organic cation, *B* is a metal cation and *X* is a halogen atom. The cation frequently used is an organic small molecule, for example methylammonium $\left(CH_3NH_3{}^+\right)$ or MA^+ formamidinium $\left(CH(NH_2)_2{}^+\right)$ or FA^+. Usually, the divalent metal cation is Pb^{2+} or Sn^{2+} and the anion *X* is Cl^-, Br^- or I^- (see Figure 2.11).

Lead halide-based perovskite, with formula $CH_3NH_3\,PbI_3$, or *MAPI*, is the most widely studied as an absorber material for solar cells. Its electrical properties are very good, with high charge carrier mobility: 20-40 $cm^2V^{-1}s^{-1}$ for holes and electrons (Leijtens *et al.* 2014). These values are modest compared to the mobilities of inorganic SCs. However, the overall efficiency of a solar cell depends largely on the lifetime of the charge carriers and their diffusion length. In perovskite, excitons are formed instantaneously and some of these excitons are dissociated in a very short length of time (2 ps). Excitons and free charges move simultaneously in the material. The effective masses of the electron and hole in perovskite are small ($m_n^* = 0.23\,m_0$ and $m_p^* = 0.29\,m_0$) (Giorgi *et al.* 2013), which helps facilitate movement of the carriers. The diffusion length of excitons (~ 100 nm) in perovskites $CH_3NH_3\,PbI_3$ is significantly greater than that in organic or polymer materials (~ 10 nm) and can even reach 1 μm in mixed perovskites $CH_3NH_3\,PbI_{3-x}Cl_x$ (Stranks *et al.* 2013). This can be explained by the particular photophysical mechanism of the material (Sum and Mathews 2014). Just after their creation, the dissociated charges and excitons move in the material and can recombine according to the mechanisms already examined (mono- or bimolecular recombination). Yet, in perovskite, these processes are inefficient or, when they occur, the recombination rate is very low, such that charges and excitons can diffuse over long distances before recombining.

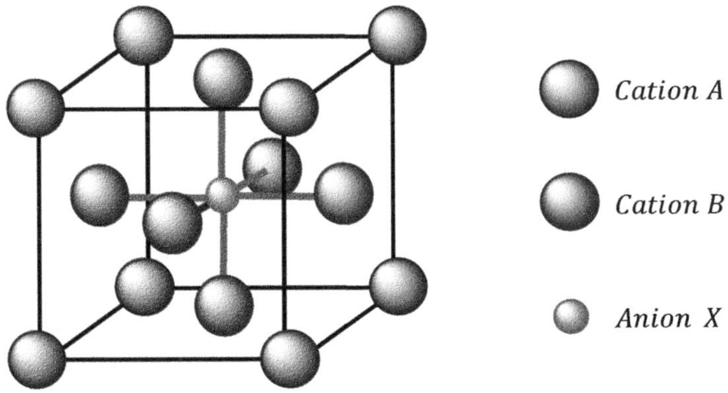

Figure 2.11. *Structure of hybrid perovskite. For a color version of this figure, see www.iste.co.uk/nguyen/electronics2.zip*

In addition to its remarkable electrical properties, perovskite has a high absorption coefficient and a narrow energy gap, between 1.50 and 1.55. The gap value is also adjustable with the material composition. By replacing iodine with a mixture of iodine and bromine, the gap can be continuously modified to optimize radiation absorption across the entire visible range.

With these good electrical and optical properties, perovskites are excellent absorbers and the efficiency of cells based on these materials is high compared to those of organic cells. According to recent results, the conversion efficiency of perovskite-based cells reaches ~22% (National Renewable Energy Laboratory 2017), which is close to that of silicon-based cells. In comparison, the production cost of perovskites is far lower than that of silicon, and perovskite solar cells are becoming of potential interest for photovoltaic devices in the near future.

2.7.1. *Structure of perovskite*

Three types of perovskite network exist. For example, consider the perovskites CH_3NH_3 PbI_3 (Sum and Mathews 2014) (see Figure 2.12):

– the 1D network consists of strings of octahedrons $(PbI_6)^{4-}$, with each octahedron connected to their two closest neighbors by the opposite lattice points. Its chemical formula is $(C_{10}H_{21}NH_3)_2PbI_4$;

– the 2D network consists of octahedron planes $(PbI_6)^{4-}$, with each octahedron connected to the four closest neighbors of the iodine ion sites. The octahedral plane is interposed between two layers of organic molecules o$CH_3NH_3{}^+$f. Its chemical formula is $(CH_3NH_3)_2PbI_4$;

– the 3D network consists of three-dimensional octahedrons $(PbI_6)^{4-}$, with each octahedron connected to the six closest neighbors of the iodine ion sites. The counter ions $CH_3NH_3{}^+$ occupy the free space between the octahedra. Its chemical formula is $CH_3NH_3PbI_3$.

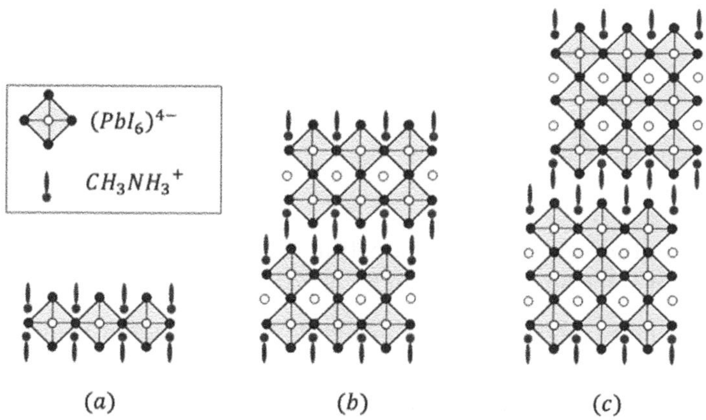

Figure 2.12. *Perovskite networks: a) 1D; b) 2D; c) 3D*

2.7.2. Solar cells based on perovskite

The realization of perovskite-based solar cells is similar to that of organic cells. The differences lie in the use of materials and in the deposition technique of perovskite films. Figure 2.13 shows the standard structure of a perovskite cell. It is composed of a transparent conducting electrode (FTO), an ETL transport layer, the perovskite absorber, an HTL transport layer and a metal electrode (Au, Ag). An N-type SC, generally titanium oxide, TiO_2, or zinc oxide, ZnO, is deposited in a mesoporous thin layer, or in nanorod arrays. The solution-deposited perovskite penetrates the mesoporous layer and forms a compact layer above this layer. The P-type SC giving the best efficiency results is *spiro-OMeTAD* or (2,2',7,7'-tetrakis-(N,N-di-4-methoxyphenylamino)-9,9'-spirobifluorene). The other possible N-type SCs used are tungsten oxide WO_3 or PCBM, and the P-type SCs are P3HT or poly(triaryl amine) (PTAA).

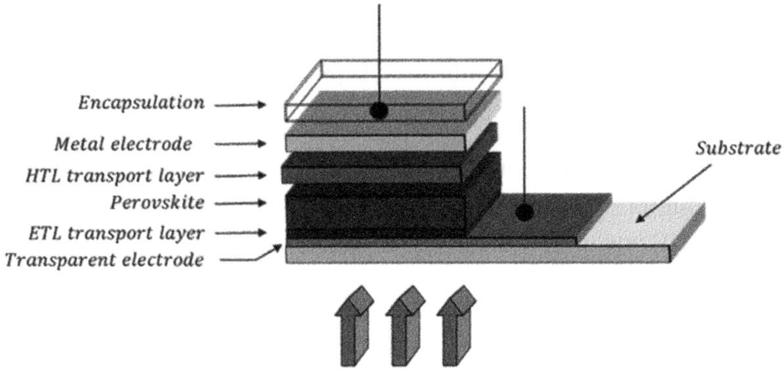

Figure 2.13. *Structure of a perovskite solar cell in standard configuration. For a color version of this figure, see www.iste.co.uk/nguyen/electronics2.zip*

The deposition of the perovskite film can be carried out using several techniques. In solution, the film can be deposited in one or two stages:

– in one stage, methylammonium iodide CH_3NH_3 I and lead iodide PbI_2 are dissolved in a solvent; N,N-dimethylformamide (DMF) or dimethylsulfoxide (DMSO) is often used. The film is deposited on the N-type SC in the standard structure by spin coating;

– in two stages, a solution of PbI_2 dissolved in DMF is used first to deposit a thin film on the N-type SC. Next, a solution of CH_3NH_3 I dissolved in isopropanol (IPA) is deposited on the previous layer. The solution penetrates the first layer and together with PbI_2 forms the perovskite. Regarding the film quality, there is no notable difference between films produced by the two methods, but the morphology of the layer is modified, which may influence the perovskite/HTM interface of the cells. On the other hand, in cells with the perovskite layer deposited in two stages, the perovskite is in full contact with the surface of the N-type SC, while with the one-stage method, the contact is incomplete.

With vacuum evaporation, the perovskite films are obtained by co-evaporation of PbI_2 and CH_3NH_3 I. The films are homogeneous and dense with average-sized grains of approximately 1 μm.

For good crystallization of the perovskite layer, annealing at 100 °C is performed after the film is deposited in order to evaporate the solvent residues, regardless of the method used (one or two stages).

2.7.3. *Conversion efficiency*

The efficiency of the first solar cells using MAPI as an absorber was 3.81% (Kojima *et al.* 2009). These cells were dye-based and used TiO_2 as an N-type SC and MAPI nanoparticles were deposited on the oxide surface. The electrolyte used contains lithium halide (LiI) and iodine dissolved in methoxyacetonitrile. Then, the use of mesoporous oxide enabled an increase in the efficiency to 6.5% (Kalinowski 2005) and, with spiro-OMeTAD to replace the liquid electrolyte, the cell efficiency became greater than 10% (Mathijssen *et al.* 2007). The structure, similar to that of organic cells, then enabled planar-type devices to be produced with standard or inverted architectures. Modifications to the MAPI material were made by replacing the cation MA^+ with formamidinium FA^+, the cation Pb with Sn, then the halogen I with Br or Cl, and these changes improved the cells' quality and performance. Using mixed perovskites such as $CH_3NH_3\ PbI_{3-x}Cl_x$ increases the open-circuit voltage and therefore increases efficiency, which reaches 16%. Partially replacing MA with FA to obtain perovskite compounds of the type $(FAPbI_3)_{1-x}(MAPbI_3)_x$ has enabled the absorption domain of the solar spectrum to be extended and the material stability to be enhanced, and the cell efficiency exceeds 20% (Pellet *et al.* 2014). Adding a small amount of cesium to the previous mixture to form a triple cation Cs/MA/FA increased efficiency to 21.1% and also cell stability and reproducibility (Saliba *et al.* 2016).

2.7.4. *Problems with the use of perovskite solar cells*

Perovskites are promising for use as absorbers in high-efficiency solar cells, especially since their high solubility allows the materials to be deposited in thin films and the cost of producing cells can therefore be reduced by their easy shaping. There are, however, three main problems posed with the use of these cells that can influence their production and marketing. These problems are lead toxicity, the operating instability of the material and hysteresis of the solar cell electrical characteristics.

2.7.4.1. *Lead toxicity*

The amount of lead in MAPI is minute, but as a matter of precaution, recycling of the material should make sure to avoid contamination of the surroundings, particularly aquatic environments, and subsequently of the human body. Lead compounds are water soluble and can be transported, resulting in the contamination of diverse biota. It is possible to replace lead with other metals, but this poses an efficiency problem, as lead plays an important role in obtaining the exceptional characteristics of perovskite.

Alternative materials can meet the solubility and high-absorption criteria and provide a perovskite of comparable quality to that of MAPI. First, elements from column IV, such as germanium (Ge) or tin (Sn), can be selected because they have an electronic configuration (Ge^{2+}, Sn^{2+}) similar to that of lead (Pb^{2+}). Cells that are CH_3NH_3 SnI_3-based have an efficiency greater than 6% (Noel *et al.* 2014) and CH_3NH_3 GeI_3-based cells have a very low efficiency of 0.20% (Krishnamoorthy *et al.* 2015). Cell performance with substituted cations is low compared to MAPI material, which can be attributed to the oxidation of Ge^{2+} and Sn^{2+} to give Ge^{4+} and Sn^{4+}. These cations are much more unstable than Pb^{2+}. Second, *alkaline earth metals* such as magnesium (Mg), calcium (Ca), strontium (Sr) and barium (Ba), or *transition metals* such as manganese (Mn), iron (Fe), cobalt (Co), nickel (Ni), etc., may also replace lead.

2.7.4.2. *Instability and degradation of perovskite*

Perovskite-based solar cells have a short lifetime because of the degradation of the active material, in this case MAPI. Three factors can trigger the degradation process in the material and the cell: oxygen and moisture, UV radiation and temperature (Niu *et al.* 2015).

Upon contact with air, oxygen and moisture, the material degrades, producing lead iodide PbI_2. For example, the reaction of MAPI with oxygen is written as:

$$2\,CH_3NH_3PbI_3 + O_2 \rightarrow 2\,PbI_2 + I_2 + 2\,CH_3NH_2 + 2\,H_2O \quad [2.17]$$

Methylammonium molecules deprotonated by dioxygen decompose by releasing lead iodide PbI_2, which is a toxic pollutant. The dark brown color of perovskite changes to light yellow after degradation, indicating the presence of lead iodide.

Similarly, water acts as a catalyst to decompose MAPI according to the reaction:

$$4\,CH_3NH_3PbI_3 + 2H_2O \rightarrow (CH_3NH_3)_4PbI_6, 2H_2O + 3\,PbI_2$$

$$(CH_3NH_3)_4PbI_6, 2H_2O + 3\,PbI_2 \rightarrow 4\,CH_3NH_3I + 4\,PbI_2 + 2H_2O \quad [2.18]$$

The performance of solar cells is degraded when exposed to UV radiation. Several experiments have suggested that titanium oxide TiO_2 is responsible for this degradation by creating deep traps that alter the transport of free charges after exciton dissociation (Leijtens *et al.* 2013).

When exposed to high temperatures, the perovskite layer is decomposed according to the reaction:

$$CH_3NH_3PbI_3 \rightarrow CH_3NH_2 + PbI_2 + HI \uparrow \quad [2.19]$$

The temperature at which the material is degraded is approximately 85°C. However, thermal degradation depends on several other factors, such as the nature of the substrate on which the perovskite layer is deposited and the environment in which the experiment is carried out. Thermal degradation of the material is significant, and temperature control of the cell in operation is required to prevent degradation. Furthermore, annealing of the material is essential in order to improve its quality, and temperature control during the operation must be rigorous to observe the limits which are not to be exceeded.

Solutions have been proposed to diminish the perovskite degradation process. Cell encapsulation is an efficient method of protecting the active layer of organic devices in general, but it can cause a loss in the mechanical flexibility that is an asset of these components.

The stability of the material can be improved by modifying the chemical structure of the perovskite. In the structure of perovskites, the organic cation is mobile and the bonds with iodine atoms are not stable. By replacing the

$CH_3NH_3^+$ cation with other, less mobile cations, such as formamidinium $CH(NH_2)_2^+$ or FA^+, the thermal stability of the material is reinforced. Other large cations such as guanidium $C(NH_2)_3^+$ or ethylammonium $C_2H_5NH_3^+$ may replace the $CH_3NH_3^+$ cation to form more stable perovskites. The constraint of cation replacement in the perovskite structure lies in the size of the atoms. If the size of the A cation is too large compared to that of the cavity formed between the octahedra BX_6, the perovskite is distorted and its optical properties are altered. For cation substitution to be efficient, the size of the A and B cations and the X anion should be chosen such that the *Goldschmidt tolerance factor, t,* is between 0.8 and 1. This factor is related to the r_A, r_B radii of the A and B cations and the radius (r_X) of the X anion by:

$$r_A + r_X = t \times \sqrt{2}\,(r_B + r_X) \quad [2.20]$$

2.7.4.3. *Hysteresis of the current–voltage characteristics*

The hysteresis of the I–V current–voltage characteristics under the illumination of perovskite solar cells is observed by a change in shape and a shift of the I–V characteristic during voltage scanning in both directions. The hysteresis measurement consists of performing a *forward scan* (FS) from the open-circuit voltage V_{OC} to zero voltage (for this voltage: $I = I_{SC}$) followed by a *reverse scan* (RS) from zero voltage ($I = I_{SC}$) to the voltage V_{OC} (see Figure 2.14). Depending on the parameters used, several variants of hysteresis measurement exist. These parameters relate to the direction, the voltage scan speed and the voltage values used.

The hysteresis effect can be quantified by the *hysteresis index* (HI), which compares the conversion efficiencies measured in both voltage scan directions:

$$HI = \frac{\eta_{RS}}{\eta_{FS}} \quad [2.21]$$

A high index indicates that the hysteresis effect is severe, whereas an index close to one means that it is negligible. Note that this definition of the index is not universal and other definitions exist, established based on studies and results (Kim and Park 2014). The index needs to be determined in order to assess the quality of the cells because it is essential to be certain of the efficiency of the cells before they can be used and marketed.

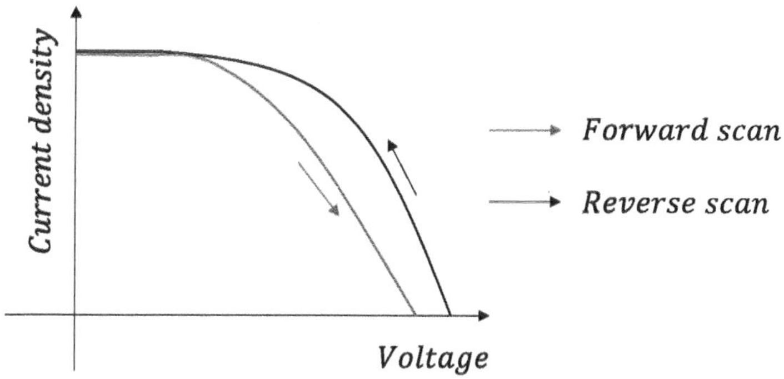

Figure 2.14. *Hysteresis effect of the current–voltage characteristic. For a color version of this figure, see www.iste.co.uk/nguyen/electronics2.zip*

Several hypotheses have been put forward to explain the origin of the hysteresis effect of perovskite solar cells: ion migration (Sacco *et al.* 2013), charge carrier trapping (Hsieh *et al.* 2018) and ferroelectric effect (Chen *et al.* 2015). In order to reduce the hysteresis effect and possibly remove it, the technique of *passivation of the surface* of contact of the transport layer can be applied in some structures (Shao *et al.* 2014). It enables elimination of the defects formed at the interface of the layers and reduces the hysteresis effect.

2.8. Solar cells based on organic, hybrid and silicon materials

The marketing of solar cells imposes constraints that manufacturers must consider and respect. These constraints were mapped by Brabec (2004) in the form of a triangle, called the *Brabec critical triangle* for photovoltaics. The vertices of this triangle represent the three requirements: efficiency, lifetime and production cost. The requirements are each of equal importance and no one requirement exceeds the others. If the manufacture of a device can satisfactorily meet two criteria, it can be considered for commercialization in a specific sector but not in the global market. This is only possible if the third criterion is also met.

A few years later, Krebs (2008) proposed his diagram, called the *Krebs diagram*, which is presented in the form of a set of three circles arranged on the vertices of an equilateral triangle. Each vertex corresponds to a similar constraint to the Brabec triangle. These constraints are *efficiency*, *process* and *stability*. The terms used are different but refer to the same criteria as those proposed by Brabec. In his diagram, Krebs specifies his criteria, namely, the cell efficiency must be greater than 10%, the stability must be greater than 10 years and the process must be that drawn up for mass production. If all of these criteria are satisfied, the device can be produced and marketed. To compare solar cells made with different organic, hybrid and inorganic materials, in Table 2.1 we give the physical characteristics of the materials, as well as the three criteria mentioned for the marketing of the devices.

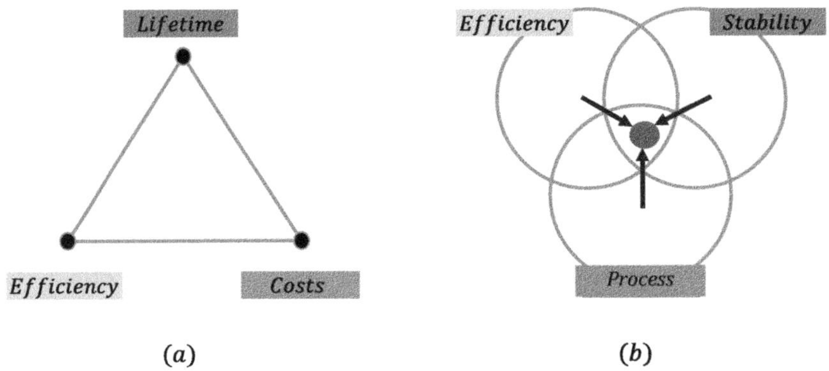

Figure 2.15. *Marketing criteria for solar cells: a) Brabec triangle; b) Krebs diagram. For a color version of this figure, see www.iste.co.uk/nguyen/electronics2.zip*

	Silicon	**Organic material**	**Hybrid material**
Structure	Crystalline	Amorphous	Crystalline
Mechanical properties	Rigid, heavy	Flexible, lightweight	Flexible, lightweight
Gap	Constant	Adjustable	Adjustable
Processing method	High temperature	Low temperature	Low temperature
Environmental impact	Non-recyclable	Low	Toxic with lead
Production cost	*High*	*Low*	*Low*
Stability	*Good*	*Low*	*Low*
Efficiency	*~25%*	*~11%*	*~21%*

Table 2.1. *Comparison of the characteristics of organic, hybrid and silicon materials for use as absorbers in solar cells*

Silicon-based cells meet two of the criteria, stability and efficiency, but the production cost is high due to the material's manufacturing process. In addition, the cells are not recyclable and pose the problem of pollution of the environment. Organic and hybrid cells meet the production cost criterion but not the stability criterion. In addition, the efficiency of organic cells is average compared to that of hybrid cells.

This table can assist in reflecting on the objective of the future development of solar cells. For the time being, silicon-based cells continue to occupy a prominent place in the global market, despite not meeting all marketing criteria satisfactorily and risking contaminating the environment if a recycling solution is not found in the near future.

Hybrid cells are a possible solution for harnessing solar energy, but it will be necessary to identify the causes of the material's instability and find solutions to enhance its stability and extend its lifetime.

As regards organic cells, the primary objective is to increase cell efficiency even if recent results show that this has exceeded the 10% threshold recommended by the Krebs diagram. Moreover, cell lifetime needs to be improved by the synthesis of new, more stable and higher-performing materials.

2.9. Strategies to improve the performance of organic and hybrid solar cells

To discuss the various strategies that can be conducted in order to improve cell performance, Figure 2.16 presents a summary of the optical and electronic processes that take place in the materials of the cell exposed to radiation.

The different stages of the photovoltaic effect have been examined and discussions regarding the formation, diffusion and dissociation of excitons have resulted in solutions to improve each process. These solutions depend mainly on the quality of the materials used in the different layers of the cell and when the choice of these materials is optimized, the cell's efficiency cannot be increased any further. The strategies to be adopted should then draw on the first stage of the photovoltaic effect, that is, the creation of excitons by absorption of photons from solar radiation.

There are two strategies to increase the absorption of photons by the material. The aim of these strategies is to broaden the domain of wavelengths of the solar spectrum absorbed by the active material. The first consists of using narrow-gap materials and the second uses multi-component or multi-junction structures.

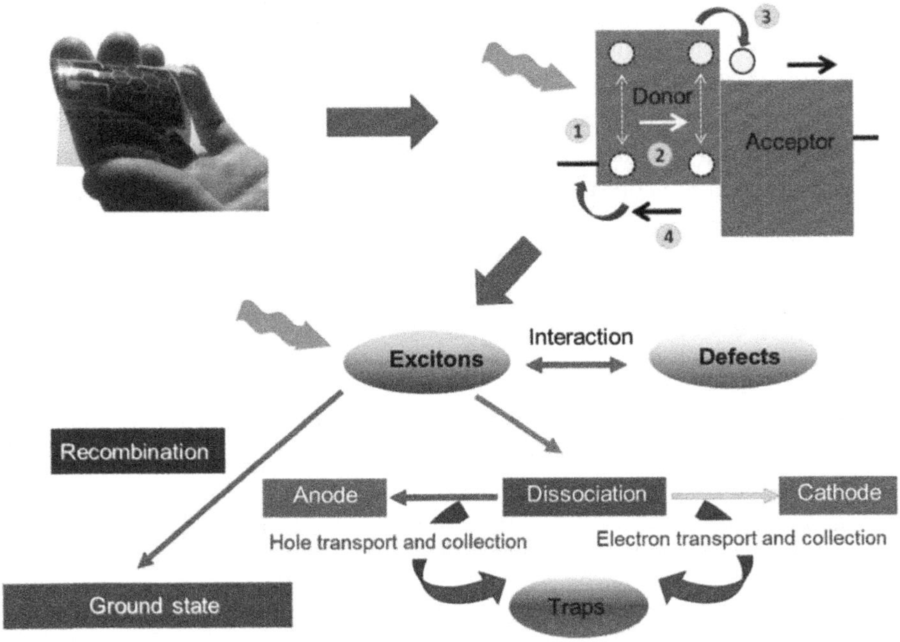

Figure 2.16. *Summary of the physical processes in an organic solar cell in operation. For a color version of this figure, see www.iste.co.uk/nguyen/electronics2.zip*

2.9.1. *Low bandgap semiconductors*

Conjugated polymers generally have a relatively wide gap, enabling them to absorb much of the solar spectrum. However, few absorb near-infrared radiation, meaning they are not very efficient when used as absorbers in solar cells. P3HT is a conjugated polymer with a gap of ~ 2 eV and is capable of absorbing photons emitted over a wavelength range of 350–650 nm. In terms of the solar spectrum on the Earth's surface, this represents approximately 46%. This means that more than half of the photons that are able to be used are not. An SC with a gap of 1.1 eV, such as silicon, would be capable of absorbing up to 77% of the radiation from the solar spectrum on the Earth's surface (Gunes *et al.* 2007). To be able to absorb photons further, organic *absorber materials* should have a narrower gap, if possible closer to that of silicon.

There are two strategies for the synthesis of low bandgap organic materials:

– the first consists of developing and using the quinoid structure of the polymer, which has a narrower gap than that of the aromatic structure. This structure is energetically unstable, however, because the energy that needs to be provided for its formation leads to the suppression of aromaticity and the structure becomes unstable. To stabilize the quinoid structure of a conjugated polymer, an aromatic ring can be incorporated into the polymer skeleton, such as poly(isothianapthene) or PITN, in which adding a benzene ring to the thiophene molecule stabilizes the quinoid form and reduces the gap to ~1 eV;

– the polymer gap can be reduced using molecules composed of *alternating donor–acceptors* (push–pull configuration). These polymers are called D–A polymers. In these polymers, the HOMO levels of the donor and acceptor interact to form two new HOMO levels. The HOMO level of the D–A monomer is principally related to the donor HOMO level (high electron concentration). Similarly, the interaction of the donor and acceptor LUMO levels creates two new LUMO levels. The LUMO level of the D–A monomer is principally related to the LUMO level of the acceptor (low electron concentration) (see Figure 2.17).

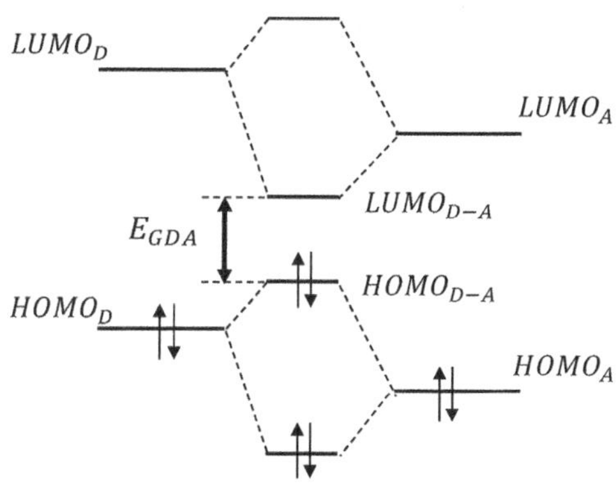

Figure 2.17. *Orbital mixing mechanism for reducing the D–A polymer gap*

Once electrons have reorganized from their original orbitals, the two new levels, $HOMO_{D\text{-}A}$ and $LUMO_{D\text{-}A}$, are created and the polymer gap is reduced with respect to the gaps of the original donors and acceptors. The range of wavelengths absorbed by D–A polymers can be extended to 800 nm, enabling the most intense radiation from the solar spectrum to be further harnessed.

D–A polymers associated with quinoid structures appear to be the most stable and efficient low bandgap absorber materials for broadening the absorption domain and increasing the conversion efficiency of solar cells accordingly. Other factors should also be taken into account in the synthesis of polymers: the flatness of the conjugated system (Roncali 1997), the solubility of the polymers (Cheng *et al.* 2009) and the possibility of their being mixed with fullerene for use in BHJ structures.

For small molecules, the same considerations apply in order to reduce the absorber gap. Materials should be soluble in common solvents.

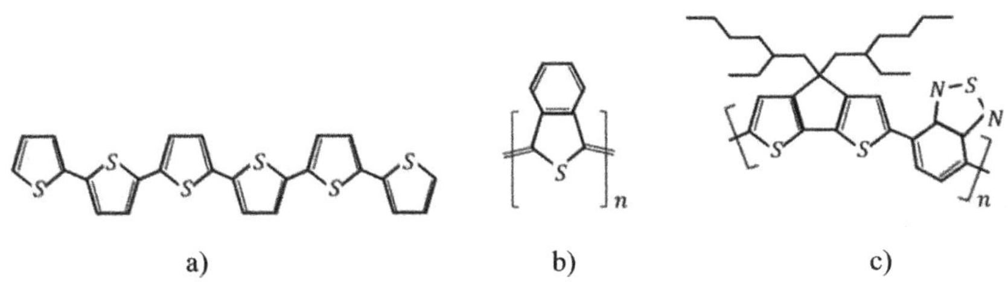

Figure 2.18. *Examples of low bandgap organic materials: a) sexithiophene (6T); b) poly(isothianapthene) (PITN); c) poly[2,6-(4,4-bis-(2-ethylhexyl)-4H-cyclopenta [2,1-b;3,4-b']dithiophene)-alt-4,7-(2,1,3-benzothiadiazole) (PCPDTBT)*

2.9.2. *Tandem cells*

To extend the range of absorbed wavelengths, normal bandgap organic materials can be used by combining them into a single solar cell, called a *tandem cell* or double-junction cell, which is composed of two *sub-cells* denoted 1 (or lower) and 2 (or higher). Each cell contains an organic material that absorbs photons in one domain of defined complementary wavelengths. The schematic diagram of a tandem cell is shown in Figure 2.19.

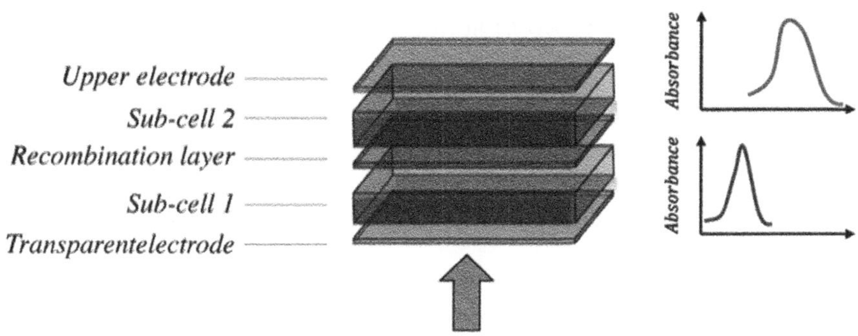

Figure 2.19. *Principle of tandem cells. For a color version of this figure, see www.iste.co.uk/nguyen/electronics2.zip*

In this figure, the tandem cell is illuminated from the cell 1 side, which filters high-energy photons. Radiation not absorbed by cell 1 is absorbed by cell 2. Between the two cells an intermediate layer is inserted, called the *recombination layer*, the role of which is to ensure the alignment of quasi-Fermi energy levels of the donor and acceptor materials of the sub-cells. This layer is semi-transparent in order to allow radiation to pass from cell 1 to cell 2. In addition, it is both an electron transport layer and a hole transport layer for both cells. In general, for tandem cells based on conjugated polymers, a mixed structure composed of a metal oxide (such as TiO_2 or ZnO) for the ETL layer and PEDOT:PSS for the HTL layer is used.

Depending on the electrical connections made on the layers, we can identify two configurations of sub-cells: sub-cells connected in series and sub-cells connected in parallel (see Figure 2.20).

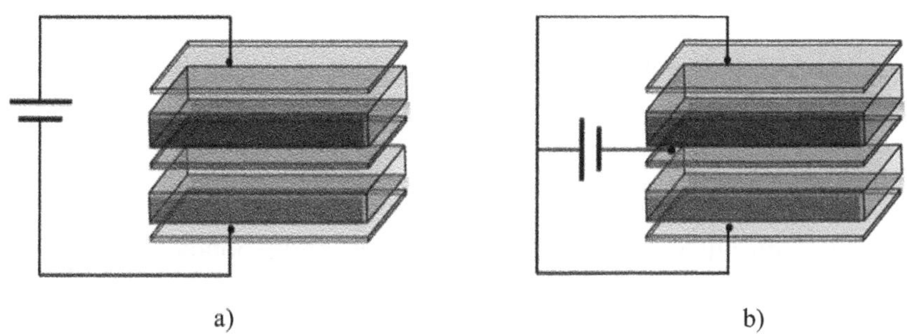

Figure 2.20. *Electrical connection of the sub-cells of a tandem cell: a) in series mode; b) in parallel mode. For a color version of this figure, see www.iste.co.uk/nguyen/electronics2.zip*

2.9.2.1. *Sub-cells connected in series or 2-terminal configuration*

In this configuration, the cells are stacked and cell 2 is deposited on cell 1. After the dissociation of excitons at the D/A interface, electrons in one sub-cell and holes in the other cell diffuse to the nearby recombination layer and recombine. The remaining free charges are collected at the electrodes of the tandem cell.

The current flowing through the tandem cell must be of the same intensity in each sub-cell. If the current in one sub-cell is greater than that generated by the other, excess charges will accumulate on contact with the recombination layer and affect the voltage at the terminals of that cell. Therefore, the total current of the tandem cell is determined by that of the lower intensity. The total open-circuit voltage is equal to the sum of the open-circuit voltages of the two sub-cells

$$V_{OC/tandem} = V_{OC1} + V_{OC2} \qquad [2.22]$$

Each of these voltages depends on the difference between the LUMO level of the acceptor and the HOMO level of the donor that make up the absorber material.

2.9.2.2. *Sub-cells connected in parallel or 3-terminal configuration*

Cells 1 and 2 have a common electrode which is the intermediate layer. The sub-cells have reversed polarities and, in operation, charges of the same type of each sub-cell are transported to the intermediate electrode. This layer plays the role of the charge collection electrode for each sub-cell. Therefore, the layer should be semi-transparent and conductive and it is common for a metal such as gold (Au) or silver (Ag) to be used.

The short-circuit current of the tandem cell is equal to the sum of the short-circuit currents of the two sub-cells:

$$J_{SC/tandem} = J_{SC1} + J_{SC2} \qquad [2.23]$$

The open-circuit voltage of the tandem cell corresponds to the lowest of the two voltages, V_{OC1} and V_{OC2}.

2.9.2.3. *Types of tandem cells depending on the material*

Materials used for active layers of sub-cells may be organic or inorganic.

Depending on their use in sub-cells, we can identify different cell types:

– tandem cells using small molecules for both sub-cells 1 and 2. The advantage of these cells is the ease of vacuum evaporation deposition of thin layers. Wide-gap donor materials may be thiophene oligomers, for example, sexithiophene (6T) or subthalocyanine chloride (SubPc). Narrow-gap acceptor materials may be zinc (ZnPc) or copper (CuPc) phthalocyanines. A metal layer is often used as an intermediate layer;

– tandem cells using conjugated polymers for the sub-cells. The advantage of these cells is the technique of film deposition by solution (spin coating or printing), which is simple and inexpensive. The disadvantage, meanwhile, is the deposition of multiple stacked layers, which requires precautions to be taken in order to avoid the new layers dissolving layers already deposited. The wide-gap polymers used are PPV and its derivatives, or P3HT with a gap of ~2 eV. One narrow-gap polymer that may be used is poly[2,6-(4,4-bis-(2-ethylhexyl)-4H-cyclopenta[2,1-b;3,4-b']dithiophene)-alt-4,7-(2,1,3-benzothiadiazole)] (PCPDTBT) with a gap of ~ 1.4 eV (Zhu *et al.* 2007). The intermediate layer deposited in solution may be PEDOT:PSS;

– mixed organic tandem cells using a conjugated polymer for sub-cell 1 (lower) and small molecules for cell 2 (upper). The conjugated polymer is deposited in solution and the small molecules by vacuum evaporation. Examples of these cells (Dennler *et al.* 2006) are those produced using a BHJ structure of P3HT:PCBM for the absorber of sub-cell 1, and through evaporation of a BHJ structure of $ZnPc : C_{60}$ for the absorber of sub-cell 2. The intermediate layer is a layer of gold that acts as a recombination layer. Tandem cells are composed of two BHJ structures. Other cells exist that are composed of several BHJs to increase conversion efficiency. They are called *multijunction cells* and have an architecture that uses materials with complementary absorption spectra to best harness the solar spectrum;

– inorganic–organic hybrid tandem cells using an inorganic absorber (Si or CIGS) for sub-cell 1 and an organic absorber for sub-cell 2. The combination of organic and inorganic materials offers two advantages. First, the use of an inorganic material helps to enhance the stability of the tandem cell. Second, photon absorption is more efficient than for organic cells because there is little or no overlap in the absorption spectra of active materials;

– one cell of particular interest is the tandem cell silicon–perovskite. The gap of silicon is ~ 1.1 eV and that of perovskite can be adjusted between 1.5

and 2.2 eV by modifying its composition. The absorption of solar radiation can therefore be optimized and the theoretical efficiency of such solar cells can exceed 30% (Filipic *et al.* 2015).

The sub-cells are connected in two configurations (see Figure 2.21). The first configuration is that of the two-wire terminal, in which one sub-cell is produced on top of the other. In order for the radiation to be absorbed by the entire cell, sub-cell 1 is based on silicon and sub-cell 2 is based on perovskite. The manufacturing process is simplified but conditions for the deposition of the layers (temperature and solvents) must be met. The second configuration is that of the four-wire terminal, in which the two cells are produced separately and then assembled together, in the same order as the two-wire terminal configuration, to form the tandem cell. Cell production takes longer and also requires strict controls of the deposit conditions. The measured conversion efficiency of silicon–perovskite-based tandem cells is greater than 26%, regardless of the configuration adopted (Bailie *et al.* 2015).

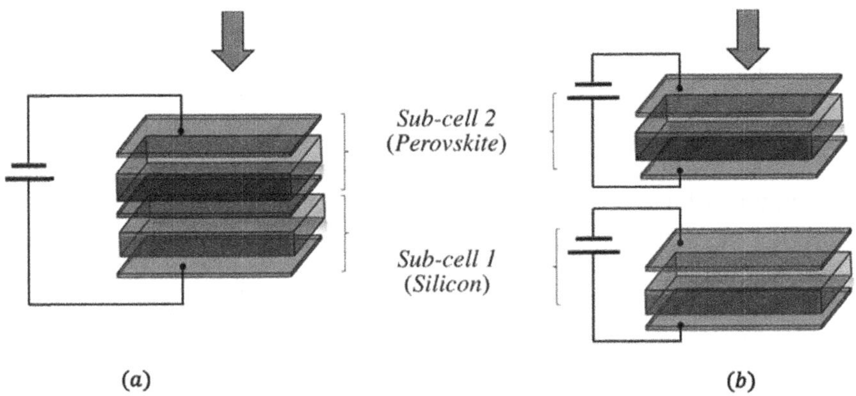

Figure 2.21. *Silicon–perovskite-based tandem cell configurations: a) two-wire terminal; b) four-wire terminal. For a color version of this figure, see www.iste.co.uk/nguyen/electronics2.zip*

2.10. Conclusion

From a strictly performance-related point of view, the conversion efficiency of organic and hybrid solar cells is greatly improved compared to that of the first cells produced in thin films. Thanks to new concepts of organic materials and cell architecture, processes of photon absorption, exciton dissociation and free-charge transport have been identified and understood, enabling the optimization of cell production parameters and better performance. The efficiency of BHJ organic cells now exceeds 10% and that of perovskite-based cells is greater than 20%. The production process has also become more efficient and effective, and mass production can potentially reduce the cell production cost. The two Brabec triangle criteria are currently satisfied, notably for perovskite-based cells. In Chapter 4, we examine the stability problems (the third criterion) of organic and hybrid devices in general and solar cells in particular, and the solutions required in order to make organic cells actually operational in the intensive harnessing of solar energy.

3

Organic Transistors

3.1. Introduction

In 1947, the invention of the *bipolar transistor* by Bardeen, Shockley and Brattain revolutionized electronic science and profoundly changed our daily lives by providing us with comforts that had been unimaginable until then. Without this revolution, we would not have experienced the development of the computers, cell phones and other devices that we use in our everyday activities today. Bipolar transistors are electronic components designed to modulate and control the current intensity flowing between two electrodes, called the *collector* and the *emitter*, by the action of a current in a third electrode, called the *base*. The same principle was adopted in 1960 to produce *field-effect transistors* (FETs), which then underwent considerable development in the design of logic circuits. In these components, the current flows between two electrodes, called the *source* and the *drain,* and its intensity is modulated using a voltage applied to a third electrode, called the *gate*. Although these components operate similarly, their physical principle and applications differ due to their physical characteristics, which we recall below.

The advent of organic electronics with the development of new materials and the acquisition of new knowledge of physical processes in this field has generated increasing interest in the study of *organic field-effect transistors* (OFETs). Indeed, because organic devices are flexible, their control circuits need to be compatible with the substrates and it would be essential for the components, particularly transistors, to be organic.

In this chapter, we study the operating principle of OFETs, the materials suitable for the characteristics of the component and the applications, both potential and already conducted, of these devices.

3.2. Operating principle

An OFET consists of several thin layers. Organic semiconductor film constitutes the active element of the component and is placed between two metal electrodes (the source and the drain). The portion of organic film limited by the edges of the electrodes S and D is called the *channel*, which has a surface area of $L \times W$.

These three layers are deposited on an insulating (dielectric) film, which insulates them from a third electrode (the gate). When a voltage, V_{DS}, is applied between the source S and the drain D, a current, I_D, called the *drain current*, flows through the organic SC. It is proportional to the SC *conductance*, g:

$$I_D = g \times V_{DS} \tag{3.1}$$

This current can be modulated by the gate voltage, V_{GS}, applied between the source S and the gate G that modifies the SC conductance through the *field effect*.

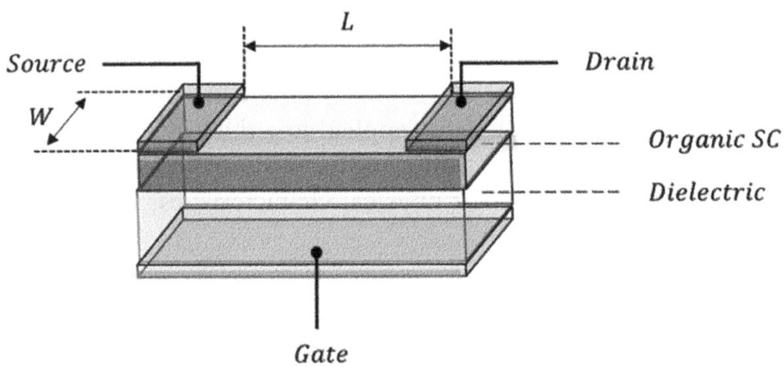

Figure 3.1. *Structure of an OFET. For a color version of this figure, see www.iste.co.uk/nguyen/electronics2.zip*

3.2.1. *Transistor effect*

A transistor consists of two junctions, P–N and N–P, with a common SC, called the *base*, that is, with an NPN or PNP structure, which are generally referred to as *NPN* or *PNP transistors*. There are several possible transistor circuit configurations and to explain the transistor effect we will consider the common emitter configuration of an NPN transistor (see Figure 3.2).

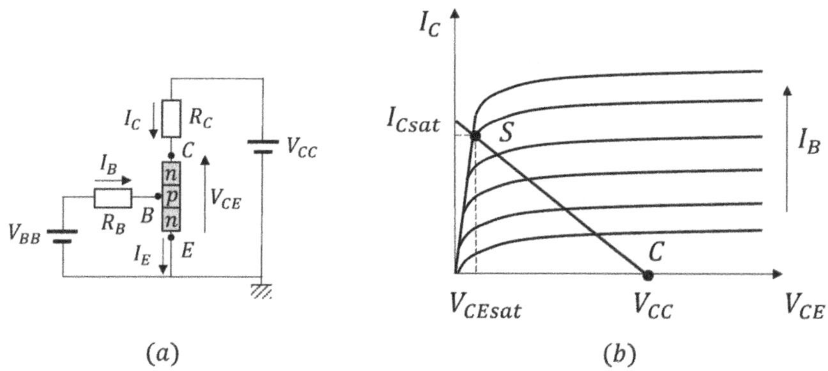

Figure 3.2. *(a) Bipolar NPN transistor in common emitter configuration; (b) current–voltage characteristics of the NPN transistor in the common emitter configuration. For a color version of this figure, see www.iste.co.uk/nguyen/electronics2.zip*

In this configuration, the base–collector (B–C) junction is reverse biased and the base–emitter (B–E) junction is forward biased. In the B–E junction, electrons are injected into the base when a voltage, V_{BB}, is applied and diffuse into the SCR of the B–C junction (see Figure 3.2(a)).

In general, the base thickness is small and the electrons can reach the collector without recombining with the majority holes of the B–C junction. For a given voltage, V_{BB}, a collector current, I_C, flows through the resistor, R_C, and a voltage, V_{CE}, is measured between the collector and the emitter.

When the base current, I_B, is very low, the collector current, I_C, is low and the potential drop in the resistor, R_C, is low. Consequently, the collector–emitter voltage, V_{CE}, is approximately equal to the bias voltage, V_{CC} (see point *C*, for cutoff, in Figure 3.2(b)). At this point, the transistor is in the cutoff mode with all the transistor currents being zero.

When the current, I_B, is very high, the collector current, I_C, becomes significant and the potential drop in the resistor, R_C, is high. As a result, the collector–emitter voltage, V_{CE}, becomes negligible (see point *S*, for saturation, in Figure 3.2(b)). At this point, the transistor is in the saturation mode with a small positive voltage, $V_{CE} = V_{CEsat}$, and a high collector current $I_C = I_{Csat}$.

Between these two limits, the collector–emitter voltage, V_{CE}, decreases as the base current, I_B, and the collector current, I_C, are increased. The current gain, β, of the common emitter configuration is the ratio between the emitter current and the base current: $\beta = I_E/I_B$. For common silicon-based transistors: $\beta > 100$.

If a sinusoidal voltage, v_B, is applied to the base of the transistor, a sinusoidal current, i_B, is superimposed onto the current, I_B, and is amplified on the collector. The output voltage, V_{CE}, comprises a sinusoidal component corresponding to the input signal, v_B, and is greatly amplified by the gain, β. The transistor operates as an *amplifier*.

A transistor also operates as a *switch* by changing the operating point, activated by the base current, I_B, passing from the saturation mode, or on-state (point *S*, $V_{CEsat} \sim 0\ V$), to the cutoff mode, or off-state (point *C*, $V_{CE} \sim V_{CC}$), and alternately returning to the saturated state. This mode of operation is common in logic electronics.

3.2.2. *Field effect*

In the OFET structure shown in Figure 3.1, we have vertical stacking of three metal–insulator–SC layers, constituting a junction, called an MIS junction. In Chapter 1, we reviewed the characteristics of a metal–SC contact and here we consider those of the MIS junction using the results acquired from metal–SC contact analysis.

3.2.2.1. *Metal–insulator–semiconductor (MIS) structures*

An MIS structure comprises a set of three layers:

– a metal layer acts as the base contact (in this case, the transistor gate, *G*);

– an insulating (or dielectric) layer of thickness *d* is inserted between the base electrode and the SC layer;

– an SC layer, assumed to be P-type here, is covered with a metal electrode for the electrical connection. This electrode is grounded in an OFET transistor.

The diagram of the energy bands of an MIS structure at equilibrium is shown in Figure 3.3(a).

Let $|e|\Phi_b$ be the difference in energy between levels E_{Fi} and E_F. We can write:

$$|e|\Phi_m = |e|\,\chi_S + \frac{E_G}{2} + |e|\,\Phi_b \quad [3.2]$$

In other words, $|e|\Phi_m = |e|\Phi_S$ and the structure is in the *flat-band regime* (see Figure 3.3(a)).

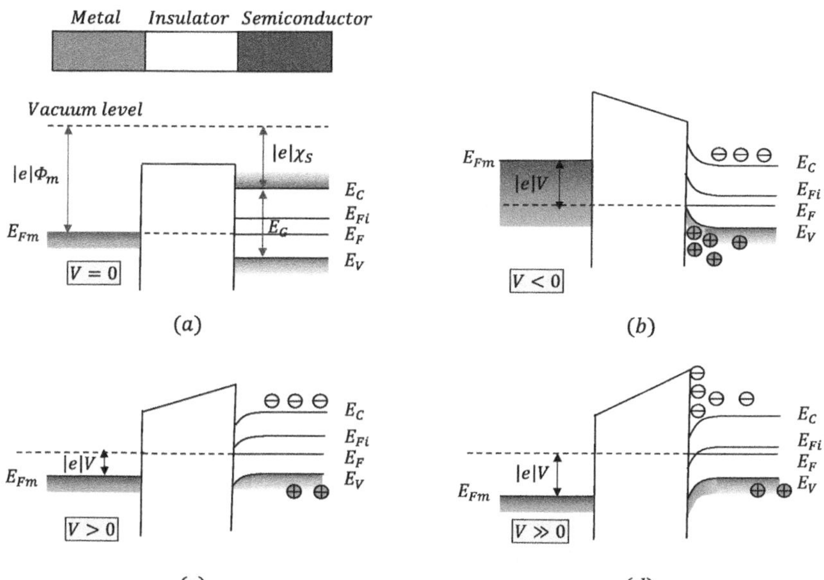

Figure 3.3. *The operating modes for MIS structures: (a) flat band; (b) accumulation; (c) depletion; (d) inversion. For a color version of this figure, see www.iste.co.uk/nguyen/electronics2.zip*

When a voltage is applied to the base electrode, the charge carriers in the SC move and modify the energy band diagram as follows:

– for a negative voltage ($V < 0$), the SC holes move towards the insulator and accumulate at the *I/S* (insulator/semiconductor) interface. Band bending occurs, raising the energy levels of the SC in the vicinity of the contact with the insulator. The Fermi level, E_F, remains constant in the SC but is closer to the VB at the interface. The structure operates in the *accumulation mode* (see Figure 3.3(b));

– for a low positive voltage, ($V > 0$), the SC holes near the I/S interface are pushed back and move away. There is a depopulation of carriers in the interface region and band bending towards the VB occurs. The Fermi level, E_F, moves away from the VB at the I/S interface. The structure operates in the *depletion mode* (see Figure 3.3(c));

– for a high positive voltage, ($V \gg 0$), the band bending becomes significant and the intrinsic Fermi level, E_{Fi}, crosses the Fermi level of the SC. As a result, the electron concentration at the interface is higher than that of the holes that are the majority carriers in the P-type SC. The structure operates in the *inversion mode* (see Figure 3.3(d)).

3.2.2.2. *Operation of OFET transistors*

Not all of the different regimes of the MIS structures are observed in organic transistors, even though these devices comprise MIS structures. This is because, in general, the organic SC is not doped and the carrier concentration is not as high as in inorganic SCs. Overall, organic transistors operate in the hole accumulation mode for the application of negative gate voltages and in the electron accumulation mode for positive gate voltages (with respect to the ground potential).

Conduction in the OFET is ensured by the current, I_{DS}, flowing between the source and the drain. Its intensity is modulated by applying the voltage, V_{GS}, to the gate electrode with respect to the ground source potential. When no voltage is applied at the transistor, it acts like an MIS capacitance and the drain current, I_{DS}, is negligible. Applying the voltage, V_{GS}, enables the conducting channel to be created by accumulation of the majority carriers (holes for a P-type SC and electrons for an N-type SC).

Let us consider the formation of the carrier channel in the case of a P-type SC. Applying a negative voltage, V_{GS}, creates an accumulation of holes at the I/S interface. On the contrary, applying a negative voltage, V_{DS}, between the (grounded) source and the (negatively biased) drain enables the flow of a hole current, I_{DS}, between these two electrodes.

When the drain voltage, V_{DS}, is low, the distribution of free carriers (holes) in the SC can be considered as uniform. The drain current intensity is proportional to the voltage applied according to Ohm's law: $I_{DS} \propto V_{DS}$. The transistor is in the *linear regime*.

As the voltage V_{DS} increases, the distribution of free carriers is modified by the dissymmetrical structure of the transistor. As the drain D is brought to a lower potential than the source, the holes accumulate at this electrode and their concentration near the source is lower. A boundary value, V_T, of V_{GS} exists, beyond which the drain current, I_{DS}, becomes constant and the transistor is in the *saturation regime*. Therefore, the voltage V_T is defined as the gate voltage, V_{GS}, that allows the transistor to pass from the depletion regime to the saturation regime. At this voltage, V_T, called the *threshold voltage*, the drain voltage V_{DS} is equal to the difference $V_{GS} - V_T$, the concentration of free carriers near the source is virtually zero and a depleted region is formed there. As a result, the conducting channel between the source and the drain is interrupted. It is said to be *pinched off*.

For each value of the gate voltage V_{GS}, the $I_{DS}(V_{DS})$ characteristic presents a linear portion followed by a constant portion. The two portions are demarked by the threshold voltage, V_T. Figure 3.4(b) shows the $I_{DS}(V_{DS})$ current–voltage characteristic of an OFET, which exhibits the same evolution as that observed in a bipolar transistor in the common emitter configuration.

The length, L, of the channel is typically from 10 to 100 μm and its width W is 10 to 1,000 times its length. To increase the transistor switching speed, the channel length, L, should be as small as possible. However, the resistance of the channel of length L and width W is given by (Bao and Locklin 2007):

$$R_{channel} = \frac{L}{W \times \mu \times Q'} \qquad [3.3]$$

where μ is the mobility of the carriers and Q' is the surface density of charges accumulated in the channel (in C/cm^2). Decreasing the length of the channel leads to an increase in its resistance, which affects the carrier transport.

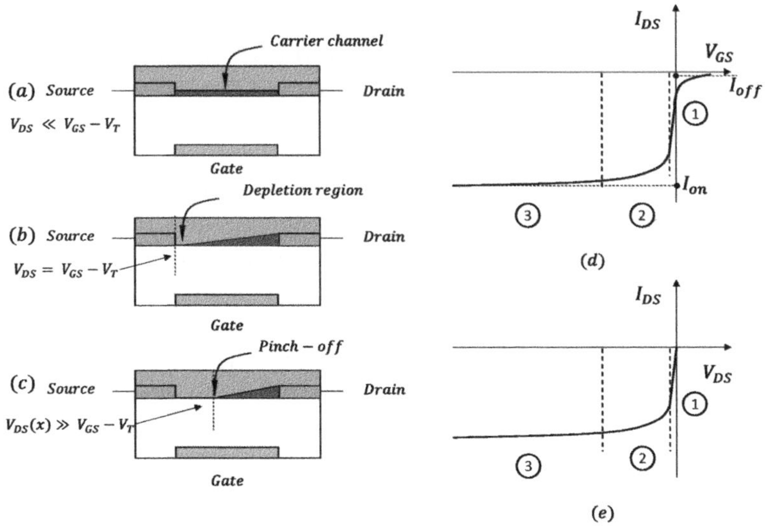

Figure 3.4. *Field effect in a p-channel OFET: (a) in the linear regime; (b) in the accumulation regime; (c) in the saturation regime; (d) $I_{DS}(V_{GS})$ transfer characteristic at constant V_{DS}; and (e) $I_{DS}(V_{DS})$ output characteristic at constant V_{GS}. For parts (d) and (e), the labels corresponding to the operating regimes are: 1 – linear regime; 2 – beginning of saturation; 3 – saturation regime. The currents I_{on} and I_{off} are shown in part (d). For a color version of this figure, see www.iste.co.uk/nguyen/electronics2.zip*

For an *n*-channel OFET, the field effect is the same as that described above. Voltages V_{GS} and V_{DS} are positive and the charge carriers in the channel are electrons.

3.2.2.3. *Structures of OFET transistors*

According to the manufacturing method, four types of OFET structure can be identified: two structures with an *upper* or *top gate* (TG) deposited at the top of the device, and two with a *lower* or *bottom gate* (BG) deposited at the bottom of the device:

– the two TG structures are divided into two substructures depending on the position of the SC layer with respect to that of the source and drain (SD) electrodes: in one substructure, the SD electrodes are above the SC, giving *upper* or *top contact* (TC), and in the other substructure, the SD electrodes are below the SC, giving *lower* or *bottom contact* (BC);

– the two BG structures are divided into two substructures, like the TG structures above. One substructure has a top contact (TC) and the other a bottom contact (BC) configuration.

The four types of structure are shown in Figure 3.5. The channel formed is indicated in dark green in the SC.

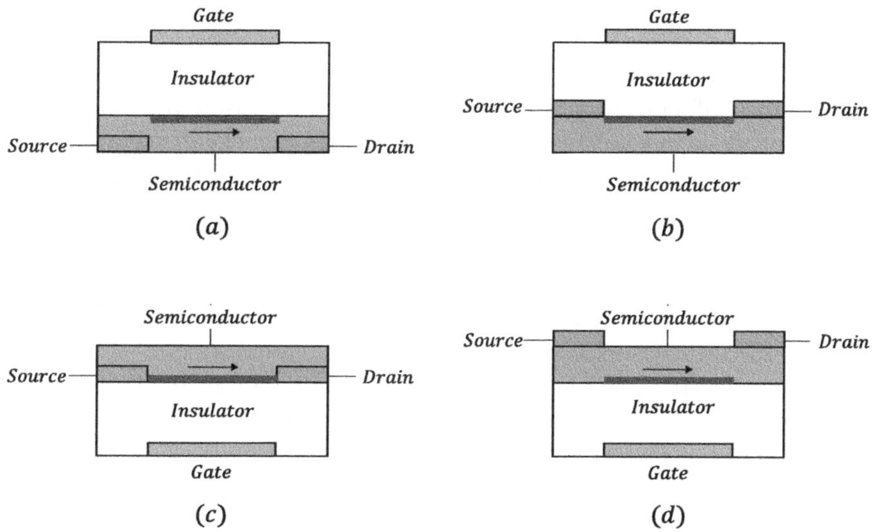

Figure 3.5. *Types of OFET structure: (a) TG/BC structure; (b) TG/TC structure; (c) BG/BC structure; (d) BG/TC structure. For a color version of this figure, see www.iste.co.uk/nguyen/electronics2.zip*

Depending on the position of the contact, the channel can directly connect the source and drain (TG/TC and BG/BC structures) or be separated from the electrodes by the SC film thickness (TG/BC and BG/TC structures). In the latter case, charge carriers must travel an additional distance in the SC volume to reach the electrodes. The TG/TC and BG/BC configurations are called *staggered configurations* and the TG/BC and BG/TC configurations are referred to as *coplanar configurations*.

Employing each type of structure presents certain advantages and disadvantages. These specific features relate to several factors, such as the contact resistance between the SC and the channel, which can become significant if the SC film is thick. The order in which the SC film is deposited by evaporation of small molecules or by polymer solution is also essential to avoid contamination or dissolution of the layers already deposited.

In general, transistors using small molecules for the SC film are manufactured with the bottom-gate (BG/TC or BG/BC) architecture, while those using a polymer film as the SC are manufactured with the top-gate (TG/TC or TG/BC) architecture.

3.2.2.4. *Current–voltage characteristics*

Two types of current–voltage characteristics are used for studying the transport process in OFETs: $I_{DS}(V_{DS})$ and $I_{DS}(V_{GS})$. The $I_{DS}(V_{DS})$ current–voltage characteristic (also called the output characteristic) can be expressed in terms of the OFET parameters, assuming that the variation in the electric field along the channel, created by V_{DS}, is smaller than that of the field perpendicular to it, created by V_{GS} (Shockley approximation) (Shur 1990). In other words, it is assumed that the thickness, d_i, of the insulating layer is negligible in comparison to the length L of the channel $d_i \ll L$, which tends to be the case for OFETs. The complete calculation is developed by considering the boundary conditions of the drain–source voltage: $x = 0$, $V(0) = 0$ and $x = L,\ V(L) = V_{DS}$ (Horowitz 1999).

In the linear regime, when $V_{DS} < V_{GS} - V_T$, the following expression of the current I_{DSL} is obtained:

$$I_{DSL} = \frac{W}{L}\,\mu_l\,C_i\,(V_{GS} - V_T)\,V_{DS} \qquad [3.4]$$

where μ_l is the carrier mobility in the linear regime, and $C_i = \frac{\varepsilon_i}{d_i}$ is the capacitance per unit area of the insulator.

The transconductance, g_m, of the channel is then:

$$g_m = \left[\frac{\delta I_{DSL}}{\delta V_{GS}}\right]_{V_{DS}=Cte} = \frac{W}{L}\,\mu_{sat}\,C_i\,V_{DS} \qquad [3.5]$$

The mobility, μ_l, can be calculated from the $I_{DS}(V_{GS})$ current–voltage characteristic (also called the transfer characteristic) by plotting I_{DSL} as a function of V_{GS}, which enables determination of μ_l by equation [3.4].

In the saturation regime, when $V_{DS} = V_{GS} - V_T$, the current I_{DSsat} is given by:

$$I_{DSsat} = \frac{W}{2\,L}\,\mu_{sat}\,C_i\,(V_{GS} - V_T)^2 \qquad [3.6]$$

That is:

$$\sqrt{I_{DSsat}} = \sqrt{\frac{W}{2\,L}\,\mu_{sat}\,C_i}\,(V_{GS} - V_T) \qquad [3.7]$$

where μ_{sat} is the carrier mobility in the saturation regime.

The transconductance, g_m, is then:

$$g_m = \left[\frac{\delta I_{DSL}}{\delta V_{GS}}\right]_{V_{DS}=Cte} = \frac{W}{L}\,\mu_{sat}\,C_i\,(V_{GS} - V_T) \qquad [3.8]$$

Likewise, the mobility μ_{sat} can be calculated by the expression [3.7], by plotting $\sqrt{I_{DSL}}$ as a function of V_{GS}. The slope of the curve provides the mobility μ_{sat} and the intercept with the axis of V_{GS} provides the threshold voltage, V_T.

In general, the mobilities determined in the saturation regime are higher than those obtained in the linear regime. The difference is explained in part by the contact resistance, which has a value comparable with that of the drain–source resistance when voltage V_{DS} is low.

3.3. Principal OFET parameters

Although OFETs are designed on the same principle as inorganic transistors, using organic SCs in place of conventional SCs alters the methods of investigation or even the nature of certain physical processes in the device. These changes are examined here in terms of specific parameters of the organic transistor. These parameters enable us to evaluate the operation of a transistor and its performance and then to look for conditions to improve them.

3.3.1. *Charge carrier mobility*

In organic SCs, the mobility of charge carriers is low and can be measured by specific techniques (time-of-flight (ToF) or CELIV) already presented in Chapter 4 of Volume 1. In the study of OFETs, carrier mobility is determined from the transfer characteristics and the values obtained may be very different from those determined by measurements performed on the same material but using the techniques mentioned above. For example, the mobility of holes in the poly(2-methoxy-5-(30:70-dimethyloctyloxy)-p-phenylene-vinylene) ($OC_1C_{10}-PP$) used as an active material in a light-emitting diode structure is $\mu_{OLED} \sim 5 \times 10^{-7} \text{cm}^2\text{V}^{-1}\text{s}^{-1}$. In an OFET structure, the mobility of holes of the same material increases to $\mu_{OFET} \sim 4.7 \times 10^{-4} \text{cm}^2\text{V}^{-1}\text{s}^{-1}$, that is, approximately three orders of magnitude higher (Tanase *et al.* 2003). The difference between these results can be attributed to the difference between carrier densities in these devices. Moreover, as we highlighted earlier, the mobility values determined by the current–voltage characteristics are different when measured in different transistor operating regimes (linear or saturation) because of a high contact resistance existing at the organic material/electrode interface. Furthermore, it has been observed that the mobility determined in the saturation regime of the transistor also depends on the value of V_{GS} and increases as the latter increases (Ryu *et al.* 2005). The possible explanation for this variation is the trapping and detrapping of charges in the SC channel, and the apparent mobility, μ_{OFET}, is given by the expression:

$$\mu_{OFET} = \mu_{free} \times \frac{Q_l}{Q_i} = \mu_{free} \times \frac{Q_l}{Q_l + Q_p} \qquad [3.9]$$

where μ_{free} is the mobility of free carriers in the extended states, Q_i is the gate-induced charge and Q_l and Q_p are the free and trapped charges in the channel, respectively.

Therefore, for device performance evaluations, the mobility values used in OFETs should be those determined from the transfer characteristics. Data on the material mobility from independent measurements will only be estimated because there may be significant deviations between these values and the effective or apparent mobility of carriers in the transistor.

3.3.2. *Contact resistance*

The interfaces at organic SC/electrode contacts influence charge injection and limit the transport of charge carriers between the source and the drain. The resistance of the interfacial region, called the *contact resistance*, R_C, determines the switching speed of the transistor and is an important parameter for the device's operation.

3.3.2.1. *Contact resistance measurement techniques*

Several techniques can be used to measure contact resistance in OFETs. Among these techniques, the transmission line technique is simple to implement and gives reliable results.

This technique consists of measuring the input resistance of a series of transistors, each of which has a determined channel length, L. In the linear regime, the input resistance, R_{ON}, measured between the source and the drain is the sum of two contributions: that of the contact, R_C, and that of the channel, R_{Ch}. The total contact resistance, R_C, is equal to the sum of the source and drain contact resistances.

$$R_{ON} = R_{Ch} + R_S + R_D = R_{Ch} + R_C$$

$$R_{ON} = \frac{L}{W\, C_i\, \mu\, (V_{GS} - V_T)} + R_C \qquad [3.10]$$

Since the resistance of the channel is proportional to its length, L, and the contact resistance is independent of it, the contact resistance is obtained by plotting the curve of the total resistance as a function of L. The contact resistance is the value of the resistance measured by extrapolation at the abscissa intercept, $L = 0$. The mobility of the carriers can also be determined by evaluating the slope of the $R_{ON}(L)$ curve.

A variation of the previous technique is the *transition-voltage technique*, which enables determination of the contact resistance using the output and transfer characteristics of a single transistor (Wang *et al.* 2010). The parameters used are the following:

– the *transition voltage,* V_{TR}, between the linear and saturation regimes of the output characteristic, $I_{DS}(V_{DS})$;

– the *saturated drain–source current,* I_{DSsat}, for a given value of the gate voltage, V_{GS}.

The source resistance, R_S, is given by:

$$R_S = \frac{\sqrt{V_{GS}-V_T}\times\left(\sqrt{2V_{TR}} - \sqrt{V_{GS}-V_T}\right)}{I_{DSsat}} \qquad [3.11]$$

The contact resistance of the transistor is $R_C = 2\,R_S$.

Other techniques exist for determining the contact resistance of OFETs, for example, the four-point method, adapted from the conventional method of measuring the conductivity of SCs (Yagi *et al.* 2004).

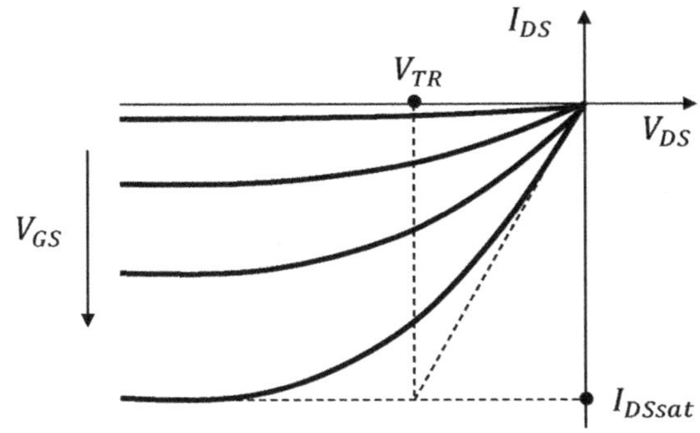

Figure 3.6. *Parameters of the transfer characteristic of a P-type OFET, used in the transition-voltage technique*

3.3.2.2. *Origin of contact resistance*

In an OFET structure, the SC is in contact with the source and drain electrodes, on the one hand, and with the insulating layer, on the other hand. The quality of each contact surface affects the transport of charges in the transistor through electrical and morphological effects. These effects were explored in Chapter 5 of Volume 1; in particular, the metal/organic SC interface was analyzed, highlighting the crucial role of charge transfer in transport at the interface. The hole or electron injection barrier is at the origin of the contact resistance at the electrode/SC interface. In this section, we study the interface formed between the SC and the insulating layer, the physical characteristics of which are different from those of the metal/SC interfaces.

3.3.2.2.1. Charge distribution at the SC/insulator contact (Horowitz 2004)

To determine the conductivity of the channel, the distribution of charges, $n(x)$, at the interface should be defined, with position variable x defined on the axis perpendicular to the interface.

With n_S the charge density at the insulator/SC interface ($x = 0$), the Poisson equation is written as:

$$\frac{d^2V}{dx^2} = -\frac{\rho(x)}{\varepsilon_S} \qquad [3.12]$$

in which $\rho(x)$ is the charge density and ε_S is the permittivity of the SC.

By applying Boltzmann's statistics to the charge distribution under consideration, we obtain:

$$\rho = |e|\, n = |e|\, n_S\, exp\left[\frac{|e|(V - V_S)}{k\,T}\right] \qquad [3.13]$$

where V_S is the potential at the insulator/SC interface ($x = 0$).

The electric field, E, corresponding to the distribution is:

$$E(x) = -\frac{dV}{dx} = \sqrt{\frac{2\,k\,Tn_S}{\varepsilon_S}}\, exp\left(\frac{|e|\,(V - V_S)}{2\,k\,T}\right) \qquad [3.14]$$

By integrating expression [3.14] on the interval $[0, x]$, we obtain:

$$\sqrt{\frac{2\,k\,Tn_S}{\varepsilon_S}}\, x = \frac{2\,k\,T}{|e|}\left[exp\left(-\frac{|e|\,(V - V_S)}{2\,k\,T}\right) - 1\right] \qquad [3.15]$$

By combining expressions [3.13] and [3.15], the expression of the charge density can be deduced:

$$n(x) = n_S\left(1 + \frac{x}{\sqrt{2}\,L_D}\right)^{-2} \qquad [3.16]$$

where L_D is the Debye length.

To calculate L_D, we calculate the integral of the expression [3.16] on the whole SC ($x \in [0, \infty]$) to obtain the total charge in the accumulation layer, which is equal to:

$$Q = C_i(V_{GS} - V_S) \sim C_i V_{GS} \quad [3.17]$$

We then deduce:

$$n(x) = \frac{(C_i V_{GS})^2}{2\, k\, T\, \varepsilon_S} \left(1 + \frac{x}{\sqrt{2}\, L_D}\right)^{-2} \quad [3.18]$$

hence:

$$L_D = \frac{\sqrt{2}\, k\, T\, \varepsilon_S}{|e|\, C_i V_{GS}} \quad [3.19]$$

Using this result, the channel can be considered as a layer of uniform thickness containing the same amount of charges as the density distribution, $n(x)$, in expression [3.16]. This thickness is equal to $\sqrt{2}\, L_D$. The Debye length is less than 1 nm in practice and the entire channel charge can be considered to be concentrated in the first layer of the SC at the interface formed with the insulator. Therefore, if the surface of the insulating layer is not perfect, morphological and electronic defects will disrupt the channel charge and, as a repercussion, the transport of charges between the source and the drain.

3.3.2.2.2. Morphological and electronic defects

The insulators used as a dielectric in OFETs are oxide or organic films. Deposited by conventional techniques, these films generally have a high surface roughness (up to 10 nm or more) and require treatments to make the surface more uniform. Other surface imperfections may be created during film shaping, for example through the incorporation of impurities, or after its formation, for example through the formation of grain boundaries.

These defects alter the potential field in which the charge carriers move and act as traps by capturing them during their transport. As a result, the surface state of the insulator has a significant influence on the operation of the device. Below we look at the materials to choose to create the insulating layer in the OFET structure.

3.3.3. *Hysteresis*

The hysteresis phenomenon is observed in OFETs in the dependence on the scan direction of voltage V_{GS} when measuring the transfer characteristic (see Figure 3.7). Depending on the scan direction (forward and reverse), the drain current varies and the voltage difference, ΔV_{hys}, between the two corresponding values of the gate voltage, V_{GS}, at the same value of the drain current, I_{DS} is used as a parameter to estimate the hysteresis. The forward direction corresponds to the voltage scan to switch the OFET on (from *off* state to *on*) and the reverse direction corresponds to the voltage scan to switch the transistor off (from *on* state to *off*).

In Figure 3.7(a), the OFET is P-type and the drain current in the reverse scan is greater at the same current in the direct scan for a given voltage, V_{GS}. The hysteresis loop is described as clockwise for this type of OFET while it is described as counterclockwise for an N-type OFET. There also exist cases where hysteresis loops are described in the opposite direction to that of the previous case, that is, the drain current in the reverse direction is less than its direct-direction value for the same V_{GS} voltage.

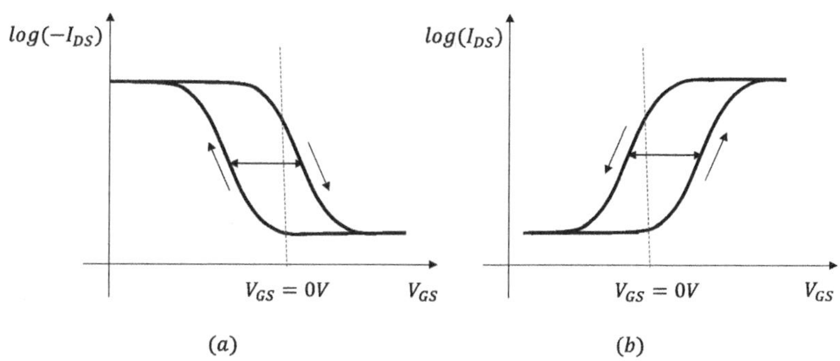

Figure 3.7. *Transfer characteristics of an OFET: (a) P-type; (b) N-type showing hysteresis*

For an OFET, hysteresis is generally attributed to the trapping of charge carriers near the channel when the loop is described in the clockwise direction, and to the displacement of mobile ions when the loop is described in the counterclockwise direction (Egginger *et al.* 2009).

3.3.4. *Gate-bias stress effects, V_{GS}*

This phenomenon corresponds to a change in the transistor threshold voltage over time when the gate voltage V_{GS} is applied and maintained for the duration of the experiment. In this case, the transistor is under gate-bias stress, which causes a shift, ΔV_T, in the threshold voltage. For a P-type OFET, the threshold voltage V_T becomes more negative than its value at the start of the stress. For an N-type OFET, V_T becomes more positive.

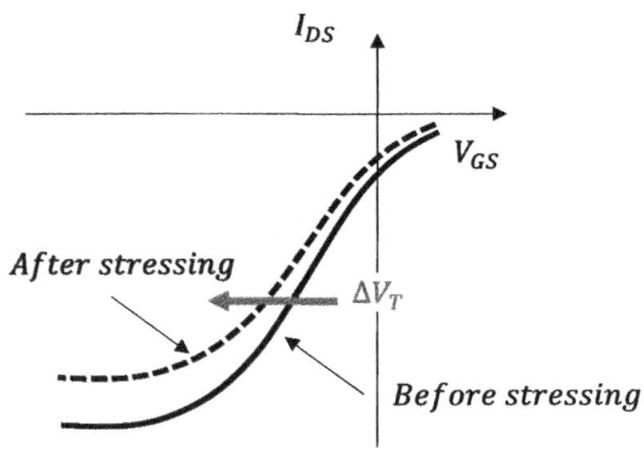

Figure 3.8. *Effect of gate-bias stress V_{GS} on the transfer characteristic*

The process is attributed to the trapping of charge carriers by traps in the SC, in the SC/insulator interface or in the insulator (Zschieschang *et al.* 2009). Charge capture increases with time, and the more the gate-bias stress is maintained, the more carriers are captured and the more the threshold voltage V_T differs from its initial value. Trapped carriers no longer participate in the transport of drain current charges and, as a result, the number of free carriers decreases and the drain current intensity, I_{DS}, decreases gradually over time. In general, when the gate-bias stress is suppressed, trapped charges are released and become mobile again, and the threshold voltage returns to its initial value before stressing.

3.3.5. *I_{on}/I_{off} current ratio*

On the $I_{DS}(V_{GS})$ transfer characteristic in Figure 3.4(d), when V_{GS} increases, the drain current I_{DS} reaches a maximum value at saturation, designated by I_{on}. On the contrary, in the absence of bias, its value is not zero but is minimal, designated as I_{off}. This current is a leakage current in the insulator, often observed in dielectrics and due to the imperfection of the material.

The I_{on}/I_{off} ratio reflects the capacity for current modulation of the OFETs. In digital logic, it represents the switching ability of the transistor, passing from the cutoff mode to the saturation mode when used as a switch in logic control circuits. For OFETs, the ratio is typically $10^5 - 10^7$.

According to the criteria for employing the transistor, other parameters can also be used to characterize OFETs. Examples include starting turn-on voltage, subthreshold slope and so on.

3.4. Materials

In an OFET, three main types of material are used: an organic SC, an insulator and metals. Other materials can also be employed to create layers that enable better mechanical strength and electrical resistance in the transistor; we will study these below.

To achieve good OFET performance, the choice of materials should be combined so that the device has the best electrical characteristics in each active part of the transistor. For the source and drain, the contacts with the SC should be ohmic to facilitate the charge transport at the SC/metal interface. For the gate, the insulator should have high dielectric strength to ensure the accumulation of charges in the channel when a voltage, V_{GS}, is applied. For transport in the organic SC, carrier mobility should be high to ensure good conduction between electrodes.

3.4.1. *Metals used for electrodes*

The choice of metal for the electrodes is based on obtaining ohmic contacts to minimize resistance at the SC/metal interface. The contact resistance, R_C, should be as low as possible and should not be higher than that of the SC.

Therefore, the metal work function should be compatible with the energy levels of the SC, that is, the HOMO level for hole injection and the LUMO level for electron injection into the SC. The HOMO level of P-type organic materials is generally between -4.9 and -5.5 eV with reference to the vacuum level. An ohmic contact can be obtained with these materials using a gold ($|e|\Phi_m = -5.1$ eV) or platinum ($|e|\Phi_m = -5.5$ eV) electrode. For an N-type SC, the LUMO level is between -3 and -4 eV with respect to the vacuum level. To obtain ohmic contact with these SCs, low work-function metals such as calcium ($|e|\Phi_m = -2.9$ eV) or silver ($|e|\Phi_m = -4.2$ eV) need to be used. These metals are unstable and oxidize easily in contact with ambient air. As a result, the device should be encapsulated if such metals are used for the electrodes.

Using metal electrodes presents two types of disadvantages: high manufacturing cost and/or instability of the SC/metal interface in air. Alternative solutions include the use of doped polymers such as PEDOT:PSS, PANI or PPy or conductive nanomaterials that can also be deposited in solution and therefore adapted to the mass-production technique (by printing or roll-to-roll processing).

3.4.2. *Dielectric materials*

The role of the insulating layer in the OFET is to create the carrier channel in the SC by accumulating charges at the SC/insulator interface when a voltage, V_{GS}, is applied to the transistor gate. In direct contact with the SC, the SC/insulator interface is an essential factor for the performance of the organic transistor (Veres *et al.* 2004).

3.4.2.1. *Capacitance of the insulating layer*

The insulating layer placed between the source (drain) and gate electrodes behaves like a flat capacitor with a capacitance of:

$$C_I = \varepsilon_i \times \frac{S}{d} \quad [3.20]$$

where ε_i is the permittivity of the insulator, $S = L \times W$ is the active surface and d is the thickness of the insulating layer. Per insulator unit area, the capacitance is $C_i = \frac{\varepsilon_i}{d}$.

The charge per insulator unit area that appears on the surface of the capacitor conductors under the voltage V_{GS} applied is:

$$Q = C_i \times V_{GS} = \frac{\varepsilon_i}{d} \times V_{GS} \qquad [3.21]$$

This charge is high if an insulator with a high permittivity and/or low thickness is used.

3.4.2.2. *Effects of the insulating layer*

The insulating layer influences the transport of carriers in the organic SC by modifying the morphology and roughness of the organic layer. Deposited on a rough insulator, the surface of the SC film has a high roughness because of the formation of large-size grain boundaries. Consequently, the mobility of the carriers is lower than that measured in the same SC deposited on a smoother insulator surface. As we have seen, the low mobility of the carriers can be explained by the high concentration of defects due to grain boundaries in this case.

When the (insulator and SC) films are deposited in solution, precautions should be taken to avoid solvent dissolution of the lower layer during deposition of the upper layer. Indeed, dissolution, even weak or partial, would result in the formation of a mixture of materials that would deform the contact surface between the films and increase the interface roughness and alter conduction in the SC (Sirringhaus 2005).

The chemical composition of the insulator may also influence the mobility of charge carriers in the organic SC. In general, electron mobility measured in OFETs is low compared to that of holes. It has been shown that when the insulator does not contain side chains such as hydroxyls, carbonyls or silanols, electron mobility in the organic SC is as high as that of holes (Chua *et al.* 2005).

In other words, conduction in the SC depends not only on the nature of the organic material but also on the insulator in contact with the SC. This dependence is attributed to potential electron trapping at the SC/insulator interface, so choosing an appropriate insulator can neutralize the traps and improve conduction in the organic SC.

3.4.2.3. *Insulating materials*

Materials used as an insulator in OFETs may be inorganic, organic or hybrid, that is, a combination of inorganic and organic materials. Employing each type of material presents certain advantages and disadvantages.

3.4.2.3.1. Inorganic insulators

These are generally wide bandgap metal oxides that have already been used in inorganic SC technology. Silicon dioxide (SiO_2) was widely used as an insulator in early OFETs with a P-type organic SC. Other materials often used as an insulator are: tantalum oxide $\left(Ta_2O_5\right)$, titanium dioxide (TiO_2), aluminum oxide $\left(Al_2O_3\right)$ and silicon nitride (Si_3N_4). Table 8.1 gives the relative permittivity, ε_r, of these materials. In principle, high permittivity of the insulator enables the amount of charges in the SC to be increased. However, in some OFETs, carrier mobility decreases as ε_r increases (Stassen *et al.* 2004) and using high-permittivity insulators is not always beneficial to the operation of the device.

Inorganic insulators have already been studied and used in conventional SC-based transistors and their characteristics are well known. They are stable, reliable and have good dielectric properties. However, being rigid, they are not compatible with organic devices and their deposition consumes a large amount of energy, making them expensive. In addition, the structure and quality of deposited films are not perfect in most cases and treatments are often required to reduce leakage currents.

Insulator	SiO_2	Ta_2O_5	TiO_2	Al_2O_3	Si_3N_4
Permittivity ε_r	3.9	25	50	9	7.8

Table 3.1. *Relative permittivity of inorganic insulators*

3.4.2.3.2. Organic insulators

By using organic insulators in an OFET, the flexibility of the device is preserved because the materials are polymers in solution and are deposited by the usual methods, such as spin coating or printing. Deposition of these materials is performed at room temperature and consumes only a little energy, and it is therefore economical.

The materials often used are non-conjugated polymers such as poly(methyl methacrylate) (PMMA), polyimide (PI), poly-4-vinylphenol (PVP), polyvinyl alcohol (PVA), cyanoethylpullulane (CYEP) and poly(perfluoro-ethylene-co-butenyl vinyl ether) (CYTOP). The relative permittivity of these materials is given in Table 3.2.

Organic insulator	PMMA	PVP	PVA	CYEP	CYTOP
Permittivity ε_r	3.5	4.5	10.4	12	2.1

Table 3.2. *Relative permittivity of organic insulators*

The permittivity values are often low and, for the current I_{DS} to be sufficient, the OFET transistor should be biased at high voltage, up to several tens of volts. Using such voltages does not facilitate the assembly of the transistor circuit. To reduce the drain voltage V_{DS} to values comparable to those used for inorganic transistors, that is, a few volts, the insulator thickness, which is of the order of a few hundred nanometers, should be reduced. However, a film that is too thin can be damaged by a high electric field, and therefore it should have high dielectric strength. The solution to this problem could be the use of the self-assembled monolayer or SAM layers already examined in Chapter 1.

The average thickness of an SAM layer is a few nanometers but, thanks to the ordered arrangement of the molecules, it forms a high-quality, high-capacitance capacitor and acts as a perfect OFET insulator. Indeed, these layers are virtually free of defects and the capacitor leakage current is almost zero. Moreover, the only slightly rough surface of the layers constitutes a perfectly flat and defect-free capacitor conductor. This benefits the transport process in the SC and enables high mobility of the carriers to be obtained. Using SAM layers in a sexithiophene-based OFET enables it to be operated with a voltage of the order of 2 volts (Collet *et al.* 2000).

The molecules frequently used for the SAM layers of OFETs are hexamethyldisilazane (HMDS) and octadecyltrichlorosilane (OTS).

3.4.2.3.3. Hybrid insulators

These are insulators formed by at least two materials combined to strengthen the gate insulation from the source and drain electrodes.

The most widely used hybrid insulator is that formed by a metal oxide with an SAM layer deposited on the oxide surface. For example, the self-assembled monolayer (hexamethyldisilazane (HMDS) or octadecyltrichlorosilane (OTS)) is used to treat the oxide surface by neutralizing the carbonyl or hydroxyl side chains formed during deposition of SiO_2. In these structures, the first self-assembled monolayer is essential because it assures the transport of charges in the transistor. The extra layers would not bring any significant improvement to the mobility, which saturates as the number of monolayers increases. Instead of SAM layers, an insulating polymer layer deposited on the metal oxide layer can also be used. The neutralization effects of oxide-layer surface defects are similar to those obtained with SAM layers. In PPV derivative-based OFETs (OC_1C_{10}–PPV, MEH–PPV), using PMMA or PVA with the oxide SiO_2 makes it possible to significantly increase the mobility of holes by passivation of surface defects (Todescato *et al.* 2008).

The hybrid insulator can be formed by stacking two insulating polymer films, the combination of which leads to an insulating layer that has a higher efficacy than the films used on their own. For example, the combination of PVA and PVP polymers in a pentacene-based OFET reduces the hysteresis process and increases carrier mobility (Park *et al.* 2004). The insulating bilayer has a low-thickness layer (PVP) in contact with the SC to improve the electrical properties and a thicker layer (PVA) providing better dielectric properties to the whole.

Finally, the hybrid insulating layer may be a composite material formed of two or more insulating materials. Typically, we use a polymer in solution, into which controlled-concentration metal oxide nanoparticles are incorporated. The deposited film is flexible and its permittivity is high thanks to the presence of nanoparticles. Moreover, permittivity can also be controlled by varying the concentration of nanoparticles. A flexible insulator with good dielectric properties is therefore produced, which adapts to the structure of the OFETs. As an example, the current I_{DS} measured in pentacene-based OFETs with the composite insulator of PVP and TiO_2 nanoparticles is higher than that obtained with the PVP insulator alone (Chen *et al.* 2004). As the concentration of TiO_2 increases, the permittivity of the composite layer increases, thus increasing the output current. The disadvantage of these composite materials lies in the difficulty of obtaining a homogeneous, uniform distribution of nanoparticle concentrations. Indeed, when the nanoparticle concentration increases, they tend to agglomerate, both forming conduction paths in the polymer matrix and modifying its roughness. These effects will cause a decrease in transistor performance.

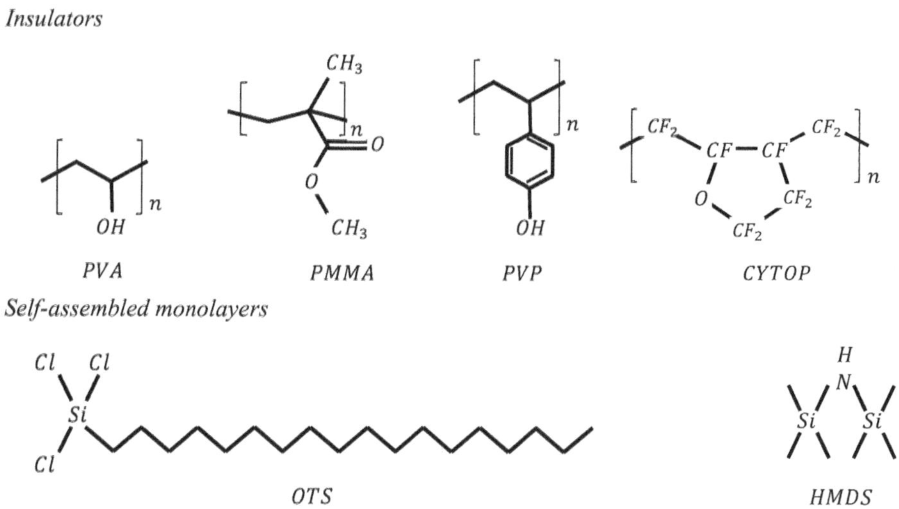

Figure 3.9. *Examples of insulating and SAM materials used in OFETs*

3.4.3. *Active organic materials*

The performance of an OFET is essentially dependent on the organic SC constituting the active layer of the device. When a voltage, V_{DS}, is applied, the drain current, I_{DS}, flowing through the SC between the *S* and *D* electrodes is measured as a function of the gate voltage, V_{GS}. To measure the drain current I_{DS} (of the order of a few tens of microamperes), it is necessary to apply a fairly large voltage, V_{DS} (of the order of a few tens of volts), because the mobility of carriers in organic materials is low. At present, measurements of mobility in OFETs show that it does not exceed 10 $cm^2V^{-1}s^{-1}$ for the best active materials.

The choice of organic SC for the active film is based primarily on carrier mobility in the material. To achieve high mobility, the material should meet the following conditions:

– the HOMO and LUMO energy levels should correspond to those of the electrodes to favor the injection and extraction of charges;

– the structure of the material should favor overlapping of the molecular orbitals to facilitate movement of the charges;

– the material should be purified and contain very few impurities to limit charge trapping;

– the organic molecules should be oriented in the direction of the substrate plane for efficient transport of charges.

Other criteria are also to be considered from both a practical and an economic perspective. On the one hand, polymer films are deposited in solution at room temperature and the manufacturing cost is lower than that of small molecules, which are generally deposited by vacuum evaporation. On the other hand, in the sequence of the deposition of films stacked by solutions, once formed, each layer should not be affected by the solvent and by the heat treatment of the subsequent layers. With these considerations and those established for electrode and dielectric materials, the main organic materials used as the SC in OFETs can be examined.

3.4.3.1. *Polymers*

In addition to the advantage due to the simple solution deposition process, polymer films normally have a fairly smooth and uniform surface with few defects to avoid or minimize interface problems with the insulator. Furthermore, printing and roll-to-roll deposition techniques can be used to make films and mass produce devices, thus reducing the manufacturing cost. These reasons explain the preferred choice of polymers as the active material in OFETs. We can identify three main conjugated polymer families:

– poly(alkylthiophene) (PAT) derivatives contain thiophene polymers with different alkyl chains, each having its own morphology and solubility. Poly(hexylthiophene), or P3HT, is one of the most widely studied polymers for OFETs. Transport in P3HT depends on the organization of the chains in the film volume, characterized by the degree of regioregularity. In a regioregular polymer, the chains form compact regions in which the skeletons' orbitals, π, overlap, enabling high carrier mobility. Moreover, the chains can organize and orient in two directions with respect to the plane of the substrate on which the film is formed: a direction perpendicular and a direction parallel to the plane. For OFETs, the chains are oriented parallel to the transistor channel and ensure good charge transport between the electrodes. The best performances of OFETs using P3HT as the SC exhibit a mobility of the order of 0.1 $cm^2V^{-1}s^{-1}$ and an I_{on}/I_{off} ratio of 10^6 (Facchetti 2007). P3HT is unstable and is easily degraded in contact with ambient air, resulting in degradation of the OFET (the current I_{off} increases). To enhance polymer stability, new polythiophene derivatives have been synthesized for use in transistors with good performance and better resistance to degradation. For example, regioregular quaterthiophene (PTQ) is used as the SC in air-operating OFETs with a mobility of 0.14 $cm^2V^{-1}s^{-1}$ and an I_{on}/I_{off} ratio of 10^7 (Ong *et al.* 2004). Similarly, measurements of transistors using poly(2,5-bis(3-alkylthiophene-2-yl)thieno(3,2-b)thiophene) (PBTTT) show a high mobility of 0.6 $cm^2V^{-1}s^{-1}$ with an I_{on}/I_{off} ratio of 10^7 (McCulloch *et al.* 2006);

– polyfluorene (PF) derivatives have good conductivity thanks to their polymer structure. In OFETs, using polyfluorene and poly(alkylthiophene) copolymers enables good electrical and stability performance to be obtained. Poly[9,9-dioctylfluorene-cobithiophene] (F8T2) is a high-performing and widely studied material for OFETs. The mobility of charge carriers in these devices reaches 0.02 $cm^2V^{-1}s^{-1}$ with an I_{on}/I_{off} ratio of 10^5 (Sirringhaus *et al.* 2000). The specific feature of this conjugated polymer is its liquid crystalline structure that allows molecules to align when deposited on an appropriate substrate. With the molecules in the ordered phase, carrier mobility increases and improves the performance of the OFET (Newsome *et al.* 2003);

– poly(phenylene-vinylene) (PPV) derivatives are widely used in OLEDs but infrequently in OFETs due to the low mobility of charge carriers. By optimizing the structure of the polymer by substitution of alkyl side chains, carrier mobility in the polymer $bisOC_{12}$–$bisC_{18}$–PPV is increased to 7 $10^{-3}cm^2V^{-1}s^{-1}$ with an I_{on}/I_{off} ratio of 10^5 and a good stability in ambient air. Better performance has been obtained with benzodifurandione-based PPV (BDPPV), in which the mobility measured reaches 1.1 $cm^2V^{-1}s^{-1}$ (Lei *et al.* 2013), a value comparable with those obtained in thiophene-based OFETs.

In most cases, the polymers used in OFETs are P-type because the hole mobility is higher than that of electrons.

One reason for this difference is the presence of electron traps at the SC/insulator interface that reduces the electron density in the channel.

However, it is important to develop N-channel OFETs to create complementary devices and circuits.

Two strategies are possible to manufacture N-type OFETs. These strategies consist of either:

– promoting the electronic conduction of polymers by substitution with electron-withdrawing groups (cyano, halogen and carbonyl) to increase the electron concentration in the channel. For example, P-type oligothiophenes are converted to N-type with fluoroalkyl substitution groups. The electron mobility measured in the OFET is then high, reaching 0.24 $cm^2V^{-1}s^{-1}$ (Facchetti *et al.* 2003);

– reducing the electron trap density of the SC/insulator interface by modifying the properties of the insulator surface.

Despite the qualities mentioned, the use of polymers in devices is subject to restrictions due to variations in the properties of industrially manufactured materials.

These variations relate to molecular mass, regioregularity and product purity. The use of small molecules can minimize these variations because these materials have well-defined molecular structures and can be synthesized with good reproducibility.

3.4.3.2. *Small molecules*

We can identify two main families of small molecules used as the SC in OFETs: acenes and thiophene-based molecules:

– acenes are a class of polycyclic aromatic hydrocarbons obtained by fusing benzene rings. Among them, *pentacene* is the most widely studied SC for OFETs. The hole mobility measured reaches 1.5 $cm^2V^{-1}s^{-1}$ with an I_{on}/I_{off} ratio greater than 10^8 (Lin *et al.* 1997). Depending on the nature of the OFET insulator, this mobility can be improved and exceeds 5 $cm^2V^{-1}s^{-1}$. However, pentacene is unstable in ambient air and absorbs visible light strongly, and these properties limit its use in practice. Among the materials of this family, *rubrene*, which is a derivative of tetracene, is a good SC for OFETs with a high mobility of 4.6 $cm^2V^{-1}s^{-1}$, close to the best values obtained with pentacene (Di *et al.* 2009);

– oligothiophenes are intensively studied for OFETs, in particular sexithiophene (6T) and its derivatives. These materials have favorable electrical and structural properties, giving high carrier mobility in the OFET. It reaches $0.002\ cm^2V^{-1}s^{-1}$ in 6T and $0.05\ cm^2V^{-1}s^{-1}$ in dihexyl-sexithiophene (DH – 6T) (Garnier *et al.* 1993).

Another type of small molecules used as active SCs in OFETs are oligophenyls.

The mobility of carriers in these materials varies from $0.01\ cm^2V^{-1}s^{-1}$ for quaterphenyl (4P) to $0.1\ cm^2V^{-1}s^{-1}$ for sexiphenyl (6P) and the I_{on}/I_{off} ratio varies from 10^5 (4P) to 10^6 (6P) (Gundlach *et al.* 1997).

One of the major disadvantages of using small molecules in OFETs is the evaporation deposition of organic materials, which consumes energy and generates costs for the production of devices.

It would be interesting to deposit small molecules by dissolving them in solution-like polymers and such a method would make it possible to use mass-production techniques, such as printing or roll-to-roll processing.

However, small molecules are poorly or not at all soluble in solvents and deposited films are fragile and easily broken because the bonding between the molecules is weak, of the van der Waals or hydrogen type.

To overcome this difficulty, synthesis techniques have been proposed and successfully applied in OFETs. These techniques are not universal; they need to be adapted to the materials used and sometimes require the transistor to have a particular configuration.

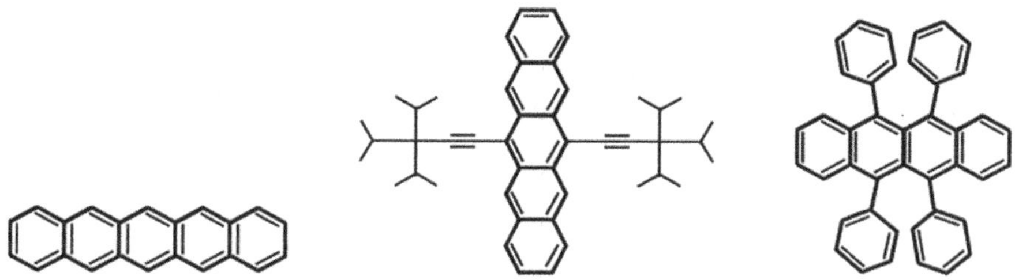

Pentacene (P5) *TIPS – Pentacene* *Rubrene*

Figure 3.10A. *Examples of active materials used in OFETs*

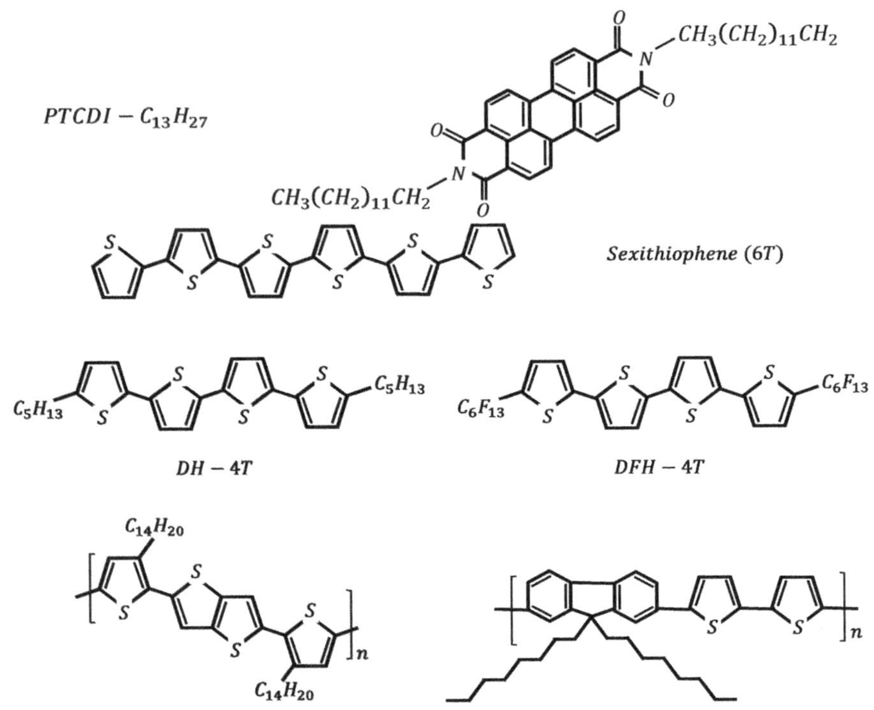

Figure 3.10B. *Examples of active materials used in OFETs (cont.)*

The first technique consists of developing a precursor that will be converted into organic material by heat conversion. For example, pentacene, obtained by converting a precursor at 180 °C, is deposited in thin film as an active layer in an OFET with good mobility (0.9 $cm^2V^{-1}s^{-1}$) and I_{on} / I_{off} (> 10^6) ratio performance (Herwig and Müllen 1999). However, this technique presents a disadvantage due to the high temperature used for conversion, which can damage previously deposited layers, including the flexible substrate. Therefore, the transistor architecture should be designed in such a way as to minimize the thermal effects of conversion.

Another approach to make small molecules soluble consists of incorporating *solubilizing groups* in their structure that increase solubility and intermolecular interactions to form crystalline domains in the material. For example, this technique is used to solubilize the molecules of 6,13-bis(triisopropyl-silylethynyl) pentacene (TIPS-pentacene) and the thin film deposited in solution in an OFET structure that works with a mobility higher than 1 $cm^2V^{-1}s^{-1}$ and an I_{on}/I_{off} ratio greater than 10^7 (Park *et al.* 2007).

3.5. Ambipolar transistors and semiconductors

Unlike inorganic SCs, which are N- or P-type by doping with donor or acceptor atoms, organic SCs are classified as N- or P-type depending on whether electrons or holes are transported preferentially. In other words, the carrier mobility determines the type of an organic SC. The classification of SCs is also based on the injection of carriers into the material. If the electrons are injected more easily than the holes, the SC is N-type and vice versa. According to this criterion, an organic SC will be N-type if its electron affinity $|e|\chi$ is high and P-type if its ionization energy is low. Therefore, the type of SC depends on the work function of the electrode, and in standard devices, the organic material behaves like an identified type of SC.

An ambipolar organic SC is defined as a material capable of transporting both electrons and holes in a device whose contacts allow the injection of two types of carriers. In other words, in an ambipolar OFET, when the applied voltage, V_{GS}, is negative, there is an accumulation of holes, and when it is positive, there will be an accumulation of electrons in the same SC.

Ambipolar OFETs work with two types of carriers, electrons and holes, and enable the realization of *organic light-emitting transistors* (OLETs). They are devices that operate as field-effect transistors and also as light-emitting diodes. The combination of functions enables the design of devices and applications of particular interest in the field of optoelectronics.

3.5.1. *Ambipolar semiconductors*

To obtain two types of carriers in an OFET, a mixture of two materials, one P-type and one N-type, can be used to form a structure of interpenetrating networks, or bulk heterojunction (BHJ), as in organic solar cells. With a suitable architecture, electrons and holes will flow through the active layer and currents are controlled by the transistor gate voltage.

It is also possible to use a single SC for the active layer in an OFET and have both types of carriers, as already explained in the study of insulators. Indeed, it is accepted that N-type conduction regime is a fairly general property of organic SCs used in OFETs, which is highlighted when high-quality insulators (without electron traps) are used. With selected materials and an appropriate transistor architecture, electron and hole currents flow through the SC and are controlled by the gate voltage. To do so, certain conditions should be met to ensure the injection of electrons and holes into the SC. Principally, the work functions of electrode metals should be compatible with the HOMO and LUMO levels of the organic SC to enable injection of two types of charges: electrons into the LUMO and holes into the HOMO bands. Therefore, two different metals should be used for the source and drain electrodes. This solution requires two separate evaporations.

Another approach consists of using a low-band gap material (< 2 eV) and one and the same metal for the electrodes such that the potential barriers at the interfaces are sufficiently low and the carriers can cross over them. These SCs have high electron affinity and low ionization energy. Moreover, the SC/insulator interface should be free of defects to prevent trapping of the carriers in the conducting channel.

As an example, OFETs have been made using either a mixture of PCBM and OC_1OC_{10}–PPV or with a single SC, poly(3,9-di-t-butylindeno[1,2-b] fluorene) (PIF) (Meijer *et al.* 2003). The output characteristics of biased transistors with positive and negative V_{GS} voltages clearly demonstrate the ambipolar character of the materials, that is, the accumulation of holes and the accumulation of electrons, depending on the bias polarity of V_{GS}.

3.5.2. *Ambipolar transistors*

An ambipolar OFET transistor can operate as an N-type transistor and as a P-type transistor depending on the applied gate voltage. Its operation can be explained as follows.

Let us assume that voltages V_{DS} and V_{GS} are positive and first consider the case of $V_{GS} = V_{DS}$. As V_{GS} is greater than $V_{t,n}$ (*n* channel threshold voltage), electrons are injected from the source and an electron current flows from the source to the drain. When V_{GS} decreases, becoming smaller than

V_{DS}, and when $V_{GS} - V_{DS} < V_{t,n}$, the source no longer injects electrons into the SC. In addition, if $V_{GS} - V_{DS} < V_{t,p}$ (p channel threshold voltage), the holes are injected through the drain and a current of holes flows to the source.

If the two preceding conditions are met ($V_{GS} > V_{t,n}$ and $V_{GS} - V_{DS} < V_{t,p}$), the electrons and holes are injected simultaneously into the SC. The transistor operates in this condition in the *ambipolar regime*. In this case, the current, I_{DS}, will not be completely saturated but presents two aspects.

When there is only one type of carrier, the $I_{DS}(V_{DS})$ characteristic is similar to that of a unipolar OFET and the current can saturate at high drain voltages, V_{DS}.

When the gate voltage, V_{GS}, enables the ambipolar regime, the current, I_{DS}, increases with V_{DS} according to a quadratic function.

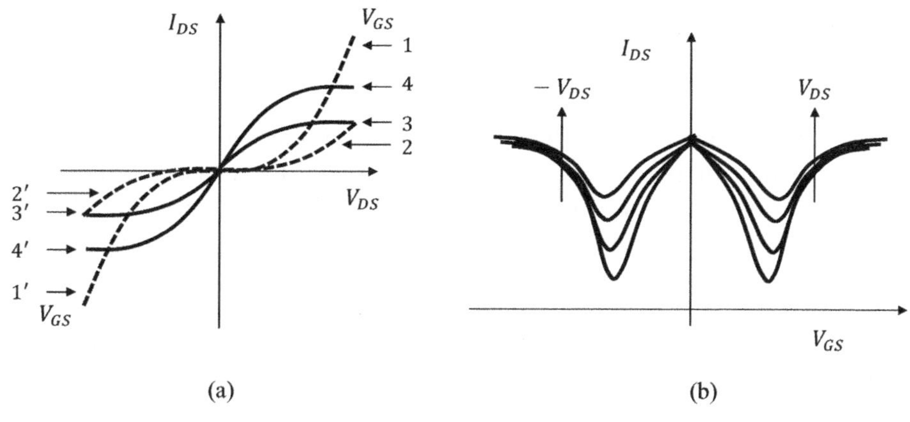

Figure 3.11. *Schematics of (a) the output characteristic and (b) the transfer characteristic of an ambipolar OFET. The curves numbered 1(1'), 2(2'), 3(3') and 4(4') correspond to the increasing values of the gate voltage, V_{GS}. Curves 1(1') and 2(2') correspond to the ambipolar regime and curves 3(3') and 4(4') correspond to the unipolar regime of the transistor*

The output and transfer characteristics comprise two networks of curves corresponding to the characteristics of the two transistors. The channels are formed when the gate voltage V_{GS} reaches the threshold voltage.

In Figure 3.11, the output and transfer characteristics of an ambipolar OFET are shown. They are symmetrical with respect to the polarities of voltages V_{DS} and V_{GS}. In reality, the mobilities of electrons and holes are not equal and the experimental curves are dissymmetrical in most cases.

3.6. Light-emitting transistors

The presence of electrons and holes in the transistor in the ambipolar regime makes it possible to consider operating it as a light-emitting diode. In addition to light emission, the transistor can also work as a switch and is becoming promising for technology using *active-matrix organic light-emitting diodes* (AMOLEDs).

3.6.1. *Ambipolar OLETs with BHJ structure*

These transistors use a mixture of two SCs forming an interpenetrating network for the active layer. Let us consider one SC with a high electron affinity, PCBM for example, and the other with a low affinity, P3HT for example. Using the composite material enables the injection and transport of charge carriers regardless of the polarity of the applied voltages. Depending on the position of the LUMO and HOMO levels, the electrons and holes remain confined in N-type or P-type materials. Consequently, the charges follow a percolation path along the channel between the electrodes. At the interface of the two materials, the electrons and holes recombine and emit photons towards the external medium. Carrier mobilities in BHJ-structured OLETs, on the contrary, are lower than those measured in transistors operating with a single SC. BHJ structures promote charge separation and therefore are better suited for solar cells than for light-emitting diodes.

3.6.2. *Single-semiconductor ambipolar OLETs*

These transistors use only one organic SC as an active layer in which the electron and hole accumulation regions are in contact. If the charge mobilities are identical, the contact region is located in the middle of the channel where the local voltage is virtually zero. In this region, electrons and holes recombine and enable light to be emitted. If the mobilities of the carriers are different, the light-emitting region moves to the drain or to the source, depending on whether the electrons or holes are more mobile.

To increase transport by improving charge balance in the OLETs, other structures have been manufactured.

Bilayer structures comprise two layers of SCs, one P-type and one N-type, deposited successively on the insulator layer of the transistor. These structures are an advantageous replacement for BHJ structures because they favor charge recombination at the interface of the two organic layers. High mobilities (3×10^{-2} $cm^{-2}V^{-1}s^{-1}$) have been obtained with this architecture using α,ω-dihexyl-quaterthiophene, or DH-4T, for the P-type SC and *N,N'*-ditridecylperylene-3,4,9,10-tetracarboxylic diimide, or PTCDI-$C_{13}H_{27}$, for the N-type SC (Dinelli *et al.* 2006). The bilayer structure can be improved by using electron and hole transport layers with the active layer.

Organic *three-layer structures* can be obtained that comprise an electron transport layer, an active layer and a hole transport layer.

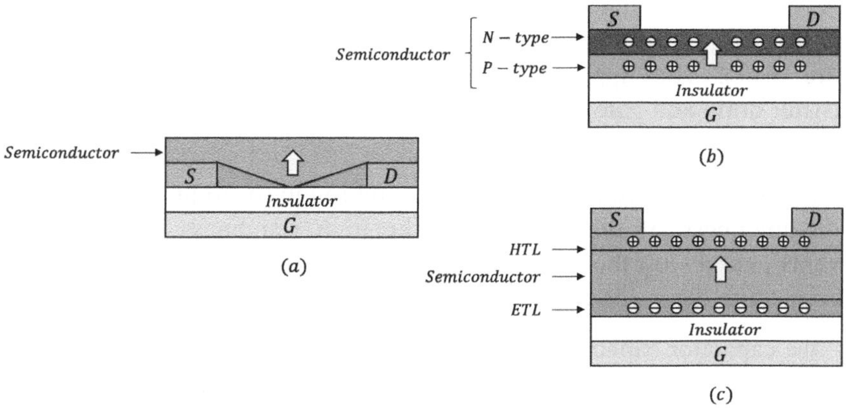

Figure 3.12. *Single-SC ambipolar OLET structures: (a) monolayer; (b) bilayer; (c) three-layer. The SC of the three-layer structure is generally a mixture of N-type and P-type SCs. For a color version of this figure, see www.iste.co.uk/nguyen/electronics2.zip*

Transport layers facilitate the injection of electrons and holes into the active material, thereby improving carrier mobilities. It is essential in these structures for the HOMO energy level of the HTL layer and the LUMO energy level of the ETL layer to be compatible with the corresponding levels of the active material. This active layer can consist of a single SC but more often consists of a mixture of N-type and P-type SCs.

An example of a three-layer OLET is that produced with an ETL layer obtained with DH-4T, an HTL layer with α,ω-diperfluorohexyl-quater thiophene, or DFH-4T, and an active layer formed by the Alq_3/Furan mixture (Capelli *et al.* 2010). The OLET's external quantum efficiency is 5%, attributed to the minimization of losses and the quenching process.

3.6.3. *Vertical OLETs*

The OFET bias voltages are of the order of several tens of volts, which is high for practical applications. To try to reduce these voltages, a new transistor structure concept is proposed. These are *vertical organic light-emitting transistors* (VOLETs), which function as a light-emitting diode and also as a switch (Ma and Yang 2004). What sets these devices apart is that they require only low V_{DS} voltages (< 10 V) to operate.

A VOLET is comprised of two parts: a capacitor and a light-emitting transistor with a common electrode, which is the source of the transistor (see Figure 3.13). The upper and lower electrodes of the VOLET are the transistor drain and gate. Two conditions are required for proper operation of the device:

– the source electrode should be thin and rough (of roughness comparable to the thickness of the electrode). The high roughness of the electrode prevents charges on the electrode surface from dissipating too quickly and, consequently, the surface of the layer is constantly charged;

– the capacitor capacitance should be high.

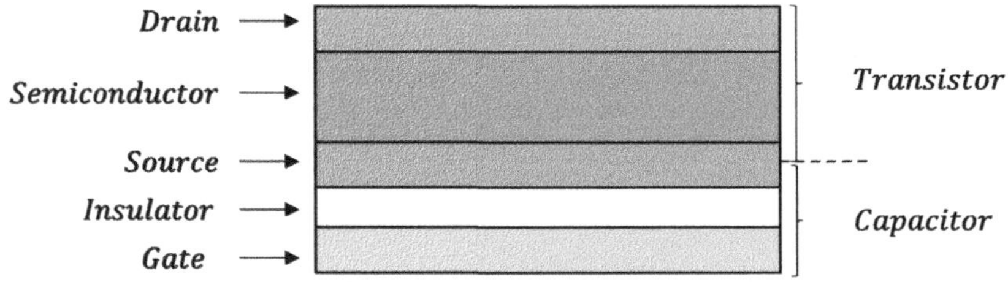

Figure 3.13. *Structure of a vertical organic light-emitting transistor (VOLET). For a color version of this figure, see www.iste.co.uk/nguyen/electronics2.zip*

Let us assume that the organic SC is N-type. The injection current is due to the injection of electrons from the source through a potential barrier, Δ_0.

When the voltage V_{GS} is not applied, the barrier Δ_0 is high and few or no electrons are injected into the SC. When a voltage $V_{GS} > 0$ is applied, the capacitor charges and the interface between the source and the insulator is negatively charged, while the interface between the source and the SC is positively charged. As a result, the initial potential barrier, Δ_0, decreases to the value $\Delta_{eff} < \Delta_0$ and electron injection into the SC increases. The drain current I_{DS} is controlled by the gate voltage and varies exponentially as a function of the potential barrier, Δ_{eff}. This leads to a high drain current, I_{DS}, for a low voltage, V_{DS}.

Using C_{60} as an organic SC, VOLETs operating with a V_{DS} voltage below 5 V can supply a current of 10 mA with a current ratio $I_{on}/I_{off} > 10^6$. These transistors are used as switches to control OLEDs.

3.7. OFET applications

Thanks to their special properties, OFETs are used to replace inorganic transistors in many applications that require these devices for signal detection, control of electronic components, memories and so on.

The advantages of OFETs over inorganic transistors lie in their simple manufacture using low-energy processes suitable for mass production (printing or roll-to-roll processing). They can also be deposited on different substrates, particularly flexible and large surfaces.

Of the many applications of OFETs, we will consider some that make use of the specific properties of organic materials and set them apart from inorganic transistors.

3.7.1. *RFID tags*

Radio frequency identification (RFID) tags are labels that are pasted or embedded onto objects that enable data to be stored and retrieved remotely for their identification. These tags are frequently used in barcode form, in electronic cards (credit cards, transport cards, identity cards, etc.) or even implanted under the skin of animals for marking.

The tag is read by a reader, which sends a signal to retrieve the data stored on the tag and supply it remotely. An antenna installed on the tag enables it to receive the signal emitted by the reader and an electronic circuit returns the data for analysis by the reader. The tag electronics mainly use silicon for the components and their miniaturization is costly, making tag manufacture expensive with silicon-based technology. Hence, OFETs are proposed for replacing silicon transistors to create RFIDs. These organic products have two essential advantages: flexibility and low price.

However, the performance of tags using OFETs is not as effective as silicon-based tags, for several reasons: the low mobility of charge carriers $\left(< 1\ \text{cm}^2\text{V}^{-1}\text{s}^{-1}\right)$, dimensions of organic transistors that are difficult to reduce to the scale of inorganic components (currently < 20 µm), high bias voltages, material instability and so on. The weak points are more difficult to control if printing processes are employed, which are the means used to mass-produce organic components and circuits. Therefore, as long as the performance of OFETs remains average for these applications, simply structured tags such as those for authentication and anti-counterfeiting can be made with organic components (Zschieschang *et al.* 2011).

3.7.2. *Sensors*

OFETs are used as sensors in numerous areas. For example, they are used as gas detectors to detect the presence of atoms, molecules and ions in the air. Such a transistor is called a *chemical-sensitive field-effect transistor* (CHEMFET). The principle of detection is based on the variation in the conductivity of the SC film when it is in contact with the analyte (or chemical species).

The use of organic materials as active SCs of transistors draws on the following advantages:

– high kinetics of adsorption and desorption processes at room temperature;

– low-energy consumption;

– detection of a wide variety of chemical compounds by modifying the structure of the organic material;

– resistance to reactive chemical compounds.

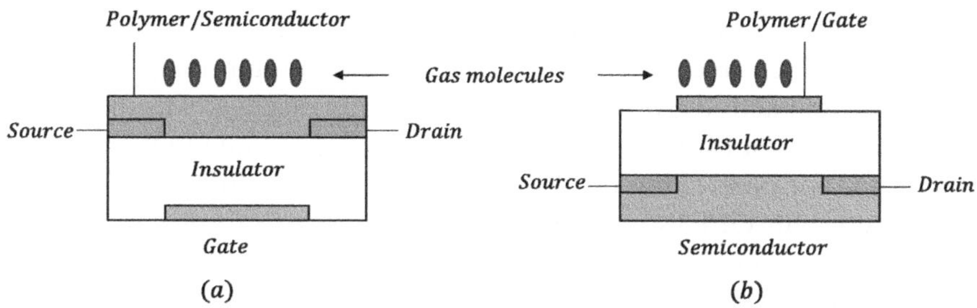

Figure 3.14. *Structure of a chemical field-effect transistor: (a) TFT structure; (b) IGFET structure. For a color version of this figure, see www.iste.co.uk/nguyen/electronics2.zip*

Two types of CHEMFET structure are used for detectors: *isolated-gate FET* (IGFET) structures and *thin-film transistor* (TFT) structures (Janata and Josowicz 2003). In IGFET structures, the organic film is the gate and is brought into contact with the gas molecules, while the SC (generally silicon) in which the field effect occurs is not exposed to gas. In TFT structures, the organic film constitutes the active SC and is exposed to gas.

As the organic film is porous, gas molecules can penetrate the surface and diffuse into the material. At the SC/insulator interface, these molecules create dipoles that alter the work function of the material at this interface and cause a shift in the threshold voltage, V_T, of the OFET. The drain current, I_{DS}, is recorded by exposing it to air containing the analyte, and current variations indicate the presence of adsorbed molecules on the surface of the SC. Concentrations of a few ppm can be detected by the sensor. The selectivity and sensitivity of detection depend on the organic SC. Various organic materials can be used for OFET sensors, such as thiophene-based polymers and oligomers, pentacene, naphthalene or copper phthalocyanine. They can detect many analytes, such as alcohols, esters, ketones and thiols (Crone *et al.* 2001).

3.7.3. *Active-matrix displays*

The low-resolution displays that equip portable electronic devices use average quality, but cost-effective, passive-matrix technology. For high-resolution displays, an active-matrix architecture is required. This component has rows and columns controlled by transistors that act as a switch to turn the screen pixels on or off. The operating principle of passive (PMOLED) and active (AMOLED) matrix OLED displays was discussed in Chapter 1. The cost of the active-matrix display is high due to the transistors incorporated for controlling the pixels. For this reason, inorganic transistors are replaced with OFETs, which have the advantage of being flexible and economical, enabling high-resolution displays to be created at low prices. In addition, using OLETs can enable replacement of the control transistor and the OLED because an OLET provides both the light-emission and switching function at each pixel. This reduces connections and simplifies the manufacturing stages of the high-resolution display and the control-circuit manufacturing cost decreases significantly (Santato *et al.* 2011).

As we have mentioned, OFET applications are numerous in various fields and it is not possible to study all of them in this chapter. Some recent developments regarding their use include phototransistors, OFET memories (Hammock *et al.* 2013), electronic skin (Guo *et al.* 2010) and so on.

3.8. Conclusion

Progress in the study of OFETs has been made in the fields of materials, physics and transistor architecture. The mobilities of polymer materials approach 1 $cm^2V^{-1}s^{-1}$ and those of small molecules reach ~ 7 $cm^2V^{-1}s^{-1}$. For organic materials they are high but, compared to the mobilities of charge carriers in inorganic SCs, they are three orders of magnitude lower. The highlighting of electronic conduction in SCs has enabled an understanding of the ambipolar character of transistors, and the realization of OLETs that emit light controlled by gate voltage to be envisaged. However, high voltages are still required for operating these devices. The concept of the vertical OFET has enabled a significant reduction in the bias voltages to values comparable with those of inorganic transistors. OFETs can be used instead of inorganic transistors for flexibility and cost benefits. To meet these criteria, the technique of printing on flexible substrates is necessary. This implies that improvements need to be made to the synthesis of materials, particularly insulating materials, in order to deposit them, free of electrical and mechanical defects, on flexible substrates. This also implies an improvement in the printing process to increase resolution without reducing the quality of the electronic conduction in the device (short-circuit or leakage current). Finally, like all other organic electronic components, the problem of material stability and device operation should also be considered in order to envisage marketing OFETs.

4

The Brabec Triangle

4.1. Introduction

In Chapter 2, we referred, in the context of the different solar cell technologies, to the diagram of the marketing criteria for devices, represented by the *Brabec critical triangle*. This triangle comprises three vertices corresponding to the conditions that any marketable product needs to satisfy to ensure a successful release. For organic solar cells, Brabec specified the following criteria: a) efficiency; b) lifetime and c) costs.

In order to enable the product to be marketed in the global market, all three criteria must be met at the same time. If one of them is not, the target market will be restricted because the product only meets the demand of a certain category of consumers. The criteria proposed by Brabec for solar cells correspond in fact to those of the consumer that each of us is when we wish to purchase any product. We ask ourselves whether the object will meet our expectations with the usual questions: is the object of good quality? Is it reasonably priced? Will it last a long time? If the answer in each case is "yes", our opinion of the product will be favorable and we will likely purchase it if we have the occasion. This reasoning leads us to extend the field of application of the Brabec triangle to all organic electronic devices, namely OLEDs, OFETs and any other organic component. In order to discuss the perspectives of organic electronics, we will examine the criteria set out for the devices studied.

In this chapter, we examine aspects of the Brabec triangle concerning light-emitting diodes, solar cells and organic transistors.

4.2. Device efficiency

4.2.1. *OLED efficiency*

OLEDs are already widely used in display applications, whether for display panels or television screens. The latest innovations concern curved or rollable OLED screens for computers or televisions. In the field of lighting, however, there are few or as yet no devices on the market, for several reasons. The efficiency of such devices still needs to be improved, whether in terms of material, charge transport or optical coupling.

To achieve high internal quantum efficiency, the emitting material should be highly fluorescent. Using phosphorescent emitters enables OLED efficiency to be improved, as singlet and triplet excitons can emit light. Thermally activated delayed fluorescent (TADF) emission can provide up to 100% internal quantum efficiency (Kaji *et al.* 2015). However, the external quantum efficiency of the diodes remains limited because the electronic conduction and optical coupling processes are not optimized yet. An important parameter for OLED efficiency is the charge balance factor, which is directly related to the difference in electron and hole mobilities. The use and selection of ETL and HTL transport layers are proving important for improving radiative recombination and, consequently, the device efficiency. In addition, it has been demonstrated that the efficiency of OLEDs decreases as the voltage applied to the diode increases (Tsutsui and Takada 2013); therefore, it is necessary to reduce the bias voltage needed for running the device. Lastly, the optical coupling factor for most emitting materials remains limited to ~ 20%, which means that 80% of photons emitted remain trapped in the diode. Therefore, optical efficiency is the main parameter to be improved in order to increase the efficiency of OLEDs. Optical losses in an OLED structure are summarized in Figure 4.1. The light emitted in the emitter needs to pass through the transparent conducting oxide (TCO) layer, then through the substrate before exiting into the air. At each interface, radiation that spreads through a layer can be entirely reflected and remain trapped in the layer. We can identify propagation modes guided in the substrate (radiation trapped between the substrate and the TCO layer) and guided in the TCO (radiation trapped between the TCO and the emitter). On the metal electrode side, the surface plasmon-polariton mode, due to oscillations of charges along the emitter/metal interface, absorbs a significant amount of energy and consequently decreases the optical efficiency of OLEDs.

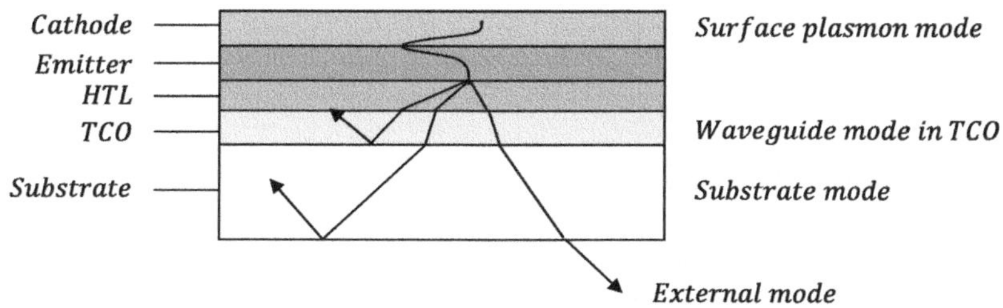

Figure 4.1. *The possible optical coupling modes in an OLED structure. For a color version of this figure, see www.iste.co.uk/nguyen/electronics2.zip*

As we saw in Chapter 1, many techniques exist to improve the extraction of light to the ambient medium. These techniques are based either on the optical characteristics of the materials or on the shape of the films to facilitate the diffusion of radiation, enabling them to escape into the external medium. For example, using microlenses or micropyramids on the emitting surface or substrate surface enables an increase in the amount of photons extracted from the diode surface (Madigan *et al.* 2000; Yamasaki *et al.* 2000). Other techniques, such as the use of corrugated surfaces (Lee *et al.* 2014), substrates with a high refractive index (Reineke *et al.* 2013) or light-scattering films (Chang *et al.* 2013), are also effective. The combination of several techniques together with a choice of high-quality materials enables efficient extraction of emitted light and high performance of the OLED. As an example, diodes based on phosphorescent materials using a substrate with a high refractive index, combined with a layer of nanoparticles for light diffusion and with a light extraction layer on the substrate surface, enable an efficacy of ~ 139 lm/W to be obtained, with a luminance of ~ 1,000 cd/m^2. At a high luminance of ~ 5,000 cd/m^2, the efficacy remains high at ~ 110 lm/W (Kato *et al.* 2015).

White organic light-emitting diodes (WOLEDs) for lighting are particularly affected by emission improvement problems because the required performance is much higher than that of diodes used in displays. Currently, WOLED panels operate with an average luminance of ~ 1,000 cd/m^2 and an average efficacy of ~ 90 lm/W (Sasabe and Kido 2013). This performance must and can be improved to meet the criteria of light sources for lighting (> 4,000 cd/m^2)

4.2.2. *Solar cell efficiency*

4.2.2.1. *Organic solar cells*

The physical processes for converting light energy to electrical energy in organic solar cells are different from those produced in inorganic cells, which influences the factors limiting efficiency in both cell types. In organic cells, the energy states are localized and the excitons generated by solar radiation are confined in small volumes, so that the energy required to dissociate the excitons is high (0.1–1.0 eV). In addition, the carrier mobilities are much lower than those measured in inorganic semiconductors (SCs), by at least three orders of magnitude, which promotes recombination. For these reasons, the power conversion efficiency of most organic cells is low compared to that of inorganic cells. The solutions proposed to improve it are the use of BHJ structures and the reduction of the film thickness to increase the transfer of charges and the collection of free charges.

The efficiency of the first organic solar cells in thin films was less than 1% (Tang 1986). The low efficiency is due to the bilayer structure used in this study, as excitons dissociate at the donor/acceptor film interface and the amount of charges collected is very low. With the concept of interpenetrating networks, cell efficiency increases because the contact surface between donor and acceptor materials increases in BHJ structures. Among the most widely used materials for these structures, P3HT:PCBM mixtures have been extensively studied and used as the active material in organic solar cells. The study of these materials has contributed significantly to the understanding of physical processes and the importance of the parameters of materials in cells for the manufacture of high-performance devices. Efficiency of P3HT:PCBM cells reached 5% in 2005 (Ma *et al.* 2005) thanks to thermal treatment, which improves film morphology and polymer crystallinity. As we mentioned in Chapter 2, P3HT:PCBM mixtures suffer from structural instability, which tends to separate phases and consequently decrease the cell efficiency. These disadvantages encourage the search for other high-quality materials and better performance for photovoltaic conversion, including low bandgap organic SCs to improve the absorption of photons from the solar spectrum. Applying the concept of D–A materials enables the SC gap to be reduced, the absorbed wavelength range to be significantly broadened and increases the cell efficiency. When the (D/A) ratio is optimized, efficiencies of more than 15% have been obtained in cells using copolymers composed of PTDB-TF, benzo[1,2-b:4,

5-b']dithiophene (BDT) and benzo[1,2-c:4,5-c']dithiophene-4,8-dione (BDD) (Cui *et al.* 2019). Such efficiencies are comparable to those of certain marketed inorganic cells and make organic cells competitive in the photovoltaic market. These efficiencies are even higher in multi-junction cells since absorption of visible light by these devices is much more efficient than light absorption of single-junction cells. As an example, high-efficiency tandems cells are produced with two cells presenting complementary absorption spectra (Meng *et al.* 2018). The upper cell uses PBDB-T poly[(2,6-(4,8-bis(5-(2-ethylhexylthio)-4-fluorothiophen-2-yl)-benzo[1,2-b:4,5-b'] dithiophene))-alt-(5,5-(1',3'-di-2-thienyl-5',7'-bis(2-ethylhexyl)benzo[1',2'-c:4',5'-c']dithiophene-4,8-dione)] as the donor material and *F-M* (thiophene-derived molecule) as the acceptor material. This cell absorbs radiation in the wavelength range of 300 to 750 nm. The lower cell uses PTB7-TH poly [4,8-bis(5-(2-ethylhexyl)thiophen-2-yl)benzo[1,2-b;4,5-b0]dithiophene-2,6-diyl-alt-(4-(2-ethylhexyl)-3-fluorothieno[3,4-b]thio-phene-)-2-carboxylate-2-6-diyl)]) as the donor material and $PC_{71}BM$ as the acceptor. This cell absorbs radiation in the wavelength range of 300 to 1,050 nm. The optimized tandem cell efficiency is 17.3%, which is comparable to cell efficiencies achieved and marketed with other technologies.

Empirical studies based on variations in donor HOMO and acceptor LUMO energy levels in BHJ structures have enabled the efficiency to be evaluated at 10–12% for single (monocomponent) cells and 15% for tandem cells (Janssen and Nelson 2013). Moreover, if we consider that the states occupied by the charges after their dissociation are different from the states occupied by the charges created by optical absorption (the difference in energy between the optical and electrical gaps is $\Delta E_{CT} = E_G - E^D_{HOMO} - E^A_{LUMO}$), the calculations lead to a conversion efficiency of 20–24%, which is comparable to the efficiency of crystalline silicon P–N junction cells. Therefore, it remains possible and feasible to improve the efficiency of actual cells. Optical and electrical losses should be optimized by minimizing:

– optical losses in inactive layers in the charge generation process;

– exciton losses by recombination, trapping by improving the quality of materials;

– losses of charges transported to the electrodes by improving charge mobilities.

4.2.2.2. *Perovskite solar cells*

The efficiency of the first perovskite solar cells was of the order of 3.8% as obtained in dye cells with a liquid electrolyte (Kojima *et al.* 2009). The perovskite used in these cells was CH_3NH_3 PbI_3. In 2012, the use of a mixed perovskite CH_3NH_3 PbI_2Cl as an absorber and spiro-OMeTAD as a hole transport layer in the cells yielded efficiency of more than 10% (Lee *et al.* 2012). Up until then, perovskite in solution was deposited on mesoporous substrates such as TiO_2 or Al_2O_3, which were assumed necessary for the proper functioning of the cells. In 2013, a new cell architecture was created by replacing the mesoporous layer of TiO_2 with a compact layer of the same oxide and depositing the layer of perovskite CH_3NH_3 $PbI_{3-x}Cl_x$ by vacuum evaporation on the oxide layer (Liu *et al.* 2013). The efficiency of these cells was 15.4%, significantly higher than that (8.6%) for cells deposited on mesoporous substrates with the same perovskite as the absorber. This study demonstrates the possibility of producing planar-structure perovskite solar cells by vacuum evaporation with high conversion efficiency. Further investigations into the role of the solvent (dimethyl sulfoxide, or DMSO) in the formation of the layer of perovskite CH_3NH_3 $Pb(I_{1-x}Br_x)_3$ resulted in efficiencies of above 18% (Jeon *et al.* 2014). The use of perovskite mixtures such as $(FAPbI_3)_{1-x}(MAPbI_3)_x$ as the absorber in planar-structure cells yields an efficiency of more than 21% (Jiang *et al.* 2017). In 2018, improvements in the manufacture of cells using the mixture $(FAPbI_3)_{0.95}(MAPbBr_3)_{0.05}$ yielded efficiencies of 23.2% (Jeon *et al.* 2018). These improvements relate to the use of N2,N2',N7,N7'-tetrakis(9,9-dimethyl-9H-fluorene-2-yl)-N2,N2',N7,N7'-tetrakis (4-methoxyphenyl)-9,9'-spirobi[fluorene]-2,2',7,7'-tetraamine or DM for the hole transport layer replacing the spiro-OMeTAD. The energy difference between the high level of the perovskite valence band and the HOMO level of DM is lower and contributes to increasing the open-circuit voltage V_{OC}, which increases the cell efficiency. The value of 23.2% is comparable with that of inorganic solar cell efficiencies and enables the consideration that perovskite technology is now potentially capable of competing with silicon cells in terms of efficiency.

Like organic cells, perovskite is used in multi-junction structures to achieve high efficiency. Tandem cells can be created by associating a perovskite solar cell with a second, low bandgap organic or inorganic cell to widen the absorption domain of the solar spectrum, with the gap of CH_3NH_3 Pb_3I being approximately 1.55 eV. Crystalline silicon is generally used as an absorber of the complementary cell, which constitutes the lower sub-cell. Near-infrared light crossing the perovskite upper sub-cell is absorbed by silicon. The thickness of the perovskite layer is chosen accordingly to enable light transmission to the lower sub-cell and at the same time to obtain sufficient photon absorption for conversion of the upper sub-cell. Moreover, for the tandem structure, using mesoporous or compact TiO_2 becomes problematic because the oxide layer requires a high annealing temperature of ~ 500°C, which may affect the silicon cell interfaces. Instead, tin oxide SnO_2 deposited at a low temperature of ~ 120°C enables the upper cell to be created without altering the performance of the silicon cell (Albrecht *et al.* 2016). The efficiency of this two-wire terminal cell is 19.9% with a short-circuit current of 14 mA.cm^{-2} and an open-circuit voltage of 1.78 V. Other cell structures have been created with different intermediate layers and similar or higher efficiencies have been obtained. By modifying the contacts between the transparent conducting film and the perovskite layer and using silver nanowires to deposit the bottom electrode of the perovskite sub-cell, the tandem cell efficiency is increased to 26.7% (Ramírez Quiroz *et al.* 2018). For efficient optical coupling of solar radiation with the gaps of the sub-cells, an optical splitter, which is a multilayer beam splitter with high reflection in the short-wavelength range and high transmission in the long-wavelength range, is added to the silicon perovskite/heterojunction tandem structure and an efficiency of 28% has been measured (Uzu *et al.* 2015).

The efficiency of perovskite tandem solar cells is found to be significantly higher than the average efficiency of currently marketed solar cells. Several calculation models predict an efficiency above 30% for these cells with optimized parameters. However, for prediction, the hypotheses used by these models are different, and it is difficult to say which of them is closest to reality. A comparative study of different sub-cell configurations and structures (parallel or series) indicates that a 45.3% efficiency would be possible for tandem perovskite/silicon cells (Futscher and Ehrler 2016).

4.2.3. *OFET performance*

For OFETs, the notion of efficiency is not appropriate, and it is preferable to speak in terms of performance. Indeed, according to the applications, OFETs are used differently and their efficiency for the function is not always evaluated by an efficiency measurement. The operation of an OFET is dependent on the mobility of the carriers, which in turn depends on the quality of the materials (SC and insulator). Several materials and their combinations enable high mobility values to be obtained, reaching several $cm^2V^{-1}s^{-1}$, and low defect densities. While these values are relatively low compared to inorganic transistors, they enable applications in various photonic and sensor fields. Furthermore, vertical structures enable a reduction in bias voltages, which are of the order of tens of volts in planar structures, to values comparable to those of silicon-based transistors for logic circuits. This performance can be further improved, but it will be difficult to predict whether they will reach the level of inorganic SCs. OFETs can be used in simple, cost-effective applications despite the current modest performance.

4.3. Stability of materials and devices

Operational stability is an essential criterion for the marketing of electronic devices, whether organic or inorganic, because a consumer would not purchase a product if they could not use it for a period of time deemed reasonable. The devices should therefore be able to operate and correctly perform the function for which they are designed during this period of time. In order to be able to effectively measure the operating time of the device, we will select a parameter that enables its function to be featured. For example, for a solar cell, the characteristic parameter is the energy conversion efficiency, whereas for an OLED, the chosen parameter is the luminance. The representative curve of the evolution of this magnitude over time enables the operating time of the device to be determined. This time is called *the lifetime,* beyond which the performance of the device is impaired or decreased compared to the performance of the device when it is switched on for the first time. The performance evaluation is based on criteria defined according to the function(s) provided by the device. The lifetime therefore varies depending on the type of electronic device and the function provided. The physical process that causes an alteration in performance is called *degradation*. It affects both the materials and the different parts of the device, including the layer interfaces. The mechanisms of degradation differ depending on the structure of the devices.

First, we describe the general processes of degradation of organic materials and devices. Then, we examine the specific processes of each type of device studied, namely OLEDs, solar cells and OFETs.

4.3.1. *Process of degradation of organic materials and devices*

Organic devices comprise films made of organic materials and/or polymers that play essential roles in their operation. Degradation of materials influences and leads to degradation of the devices. Of course, the degradation processes of isolated materials are not necessarily the same when they are incorporated into an electronic device, because of interactions due to contact with other materials used in the device. Understanding material degradation is, however, useful to study and enables material-related processes to be differentiated from those related to the structure and operation of the device.

4.3.1.1. *Material degradation*

Organic and polymer materials degrade by several processes depending on the conditions of the medium with which they are in contact or in which they are exposed. The most frequently encountered mechanisms are as follows:

– thermal degradation, which occurs during synthesis or when the material is brought to high temperature;

– chemical degradation, generated by chemical reactions with other chemical species of the external medium that alter the structure of the material, resulting in polymer chain scissions (*depolymerization reaction*) and oxidation of the material. For conjugated polymers, there is a loss in chain conjugation length;

– mechanical degradation due to the forces or stresses applied to the material causing chain scission;

– photo-degradation generated by exposure of the material to visible or UV light. The energy absorbed by the material produces reactions of the material or the impurities it contains and causes an alteration of the material structure. The action of light in the presence of oxygen in the material produces chain scissions and the formation of carbonyl groups and creates radicals that degrade the material.

Degradation is generally manifested by changes in physical properties (loss or reduction of molecular mass), mechanical properties (cracking and embrittlement) or chemical properties (change in structure and formation of functional groups: carbonyls, hydroxyls or hydroperoxides). To study changes in properties and identify degradation mechanisms, the degraded material is analyzed using optical methods (infrared, Raman and photoluminescence), surface spectroscopies (XPS, scanning electron (SEM) and atomic force (AFM) microscopy) and chemical analyses (composition and electron paramagnetic resonance (EPR)).

For illustration purposes, the study of thermo- and photooxidation of P3HT has shown that the polymer characteristic groups are degraded and lead to the formation of carbonyl derivatives and oxidized sulfur products (Manceau *et al.* 2009). The degraded alkyl chains generate radicals that are then oxidized to carboxylic acids, while the altered thiophene rings form sulfoxides, then sulfones, and finally sulfonic esters. Oxygen plays an essential role in the polymer degradation mechanism and, in its absence, the process is slowed down and decreases in intensity.

A broader study of conjugated polymers has established a number of rules regarding the structure of materials that may be degraded in ambient air and under sunlight. The degradation of these materials is promoted by the presence of side chains, cleavable bonds and exocyclic double bonds.

4.3.1.2. *Degradation of devices*

The study of the degradation of organic devices is more complex because of their architecture, which can be basic (containing a single active layer) or multilayer (containing several layers: active, transport and intermediate, for example). It is clear that studying the degradation processes of a basic device is simpler than studying a multilayer device because the degradation in the latter may originate from several parts of the device structure. In this case, it is necessary to vary several experimental parameters in order to locate and identify the mechanisms. We will not examine all of the possible degradation reactions in structures and components but only the common processes observed in organic devices, regardless of the function that they provide. Each category of device is more concerned with a particular degradation process because of their specific operating conditions. For example, solar cells, which are constantly exposed to solar radiation, may be more thermo- and photo-degraded than OLEDs.

4.3.2. *Classification of device degradation mechanisms*

Degradation mechanisms can be classified into three categories: degradation by black-spot growth, intrinsic degradation and extrinsic degradation.

4.3.2.1. *Degradation by black-spot growth*

The first degradation studies of organic devices were conducted on OLEDs with a lifetime of a few minutes. Microscopic observation of the surface of the diodes reveals concentric regions, black in the center and increasingly lighter as the diameter of the circles increases.

These regions are called *black spots (or dark spots)* and are observed interchangeably in small molecule- or polymer-based diodes (Lim *et al.* 2001). These areas correspond to the non-emissive regions of the diode and their number increases with the operating time until the light emitted is fully off. The observed circles may evolve into domes or bubbles. They can also be detached from the structure (*delamination* phenomenon).

The mechanism of black spot formation is explained by the presence of defects and impurities in the surface of the substrate, the active layer and the electrode, which promote the diffusion of oxygen or water into the device. The deformation of the cathode layer in the form of domes or bubbles would be due to the release of gas generated by oxidation and electrochemical decomposition reactions of the organic material. As the black spot regions form, the diode's emissive surface decreases and the current intensity increases, causing this surface to increase in temperature and gradually degrade.

Black spot analyses indicate a high atomic concentration of oxygen, suggesting that oxidation of the electrode is probably the degradation reaction.

Other hypotheses regarding black spot formation have also been proposed, for example, delamination of the electrode without a chemical reaction (Liew *et al.* 2000) or the formation of traps and non-radiative recombination centers in the active film by diffusion of water molecules or oxygen through the perforations in the electrode (Steiger *et al.* 2001). However, the hypothesis regarding moistures and/or oxygen diffusion seems most likely, as encapsulated devices do not exhibit such degradation in operation.

4.3.2.2. *Intrinsic degradation*

The *intrinsic degradation* of a device is manifested by a slow and continuous loss of the performance characteristic parameter as a function of the component's operating time. All external parameters (temperature, pressure, illuminance, etc.) are kept constant so that the effects produced can only originate from inside the device. Two groups of intrinsic degradation processes can be identified: electronic processes and chemical reactions.

4.3.2.2.1. Electronic processes

Degradation occurs due to the effects of charge transport during operation of the device. Depending on the architecture of the component, the nature of the organic materials and electrodes, several electronic mechanisms may degrade the device. We can identify:

– the diffusion of ions in the active layer that act as non-radiative recombination centers and affect the processes. Most devices use ITO for the transparent electrode, but this material is often unstable and can decompose on contact with active or transport layers. Decomposition releases indium and tin ions that migrate through the interfaces and enter the active layer (Nguyen *et al.* 2001);

– the transport of charge carriers in the active layer that can interact with the material, modify its electronic structure and cause irreversible degradation. As an example, this mechanism is observed in OLEDs using Alq_3 as an active layer in which a current of holes causes the ionization of molecules and transforms them into $Alq_{3cations}$ (Aziz 1999). Moreover, an accumulation of charge carriers in the vicinity of the interfaces can affect the performance and lifetime of a device by promoting reactions with molecules in that region through the high concentration of thermal and electrical energy. A charge balance in the emitter enables the recombination region to be widened and the degradation of interfacial regions to be avoided (Lee *et al.* 2005);

– trapping of charge carriers by localized states in the volume of the active material or at the interface of the layers. This process is particularly significant in OFETs and creates the shift in the threshold voltage of the transfer characteristic.

4.3.2.2.2. Chemical processes

The chemical reactions that occur in the active material modify its electronic structure, and consequently its properties, and degrade the performance of the device. There are two types of reactions described below.

Photochemical reactions

When the device is exposed to light, the excitons formed in the active material play an important role in the degradation of the device. Depending on the structure studied, degradation may occur at the ITO/organic layer interface, or in the volume of the active material, or with materials in contact with the active layer.

In the first case, the process of the material's degradation is initiated by the absorption of photons that are transferred to the ITO electrode. The oxygen molecules of the oxide, activated by the energy transfer, are released in turn, causing oxidation of the organic material.

In the second case, degradation occurs in the volume of material exposed to light in the presence or absence of oxygen. The nature of the degradation mechanism is not completely clear, and polymerization, phase transformation or oxygen diffusion reactions may occur depending on the nature of the material and the architecture of the device used.

In the third case, the degradation of the material is due to the formation of defects related to *carbonyl groups*. The formation of these defects can occur in two ways:

– an alkyl chain of a photon-excited molecule is released and a nearby oxygen molecule reacts with the excited molecule to form a ketone;

– the transfer of charges from the excited molecule (essentially triplet states) to an oxygen molecule creates a metastable state called *singlet oxygen*, which is particularly active.

This oxygen reacts with the vinyl groups of the polymer and forms different varieties of oxide in the material (Cumpston *et al.* 1997). Consequently, the oxidation of the material leads to the formation of traps, which capture charge carriers and decrease their mobility. In addition, oxygen reacts with excitons and reduces the number of radiative recombinations in OLEDs and the number of dissociated charges in solar cells.

Electrochemical reactions

These are chemical reactions that take place near the interfaces and are initiated by the charge carriers. They concern the three sections of a device: the active layer, the anode and the cathode.

In the active layer, the reaction of the organic material with the charge carriers modifies and transforms the molecules into ions as in the degradation mechanism of Alq_3, cited as an example in the electronic degradation processes.

On the anode side (generally the ITO), the residual water behaves like an electrolyte and decomposes the oxide and the part of the organic material in contact with the ITO. Water decomposition releases the indium and tin ions that diffuse into the active layer, as we described in the diffusion process. The use of PEDOT:PSS as a transport layer is particularly conducive to electrochemical degradation as this product is soluble in water. A poorly executed annealing of PEDOT:PSS leaves traces of water on the ITO surface and degrades the oxide/organic material interface and subsequently the device (Nguyen and Renaud 2009).

On the cathode side, oxidation of metals with a low work function (calcium, barium or magnesium) may be visible to the naked eye (change in appearance and/or color). For some metals of high work function (aluminum or silver), oxidation can also occur, manifested by an increase in the potential barrier at the metal/organic material interface, and leads to loss of luminance in OLEDs or an increase in series resistance in organic solar cells.

4.3.2.3. *Extrinsic degradation*

Extrinsic degradation is caused by the parameters and operating conditions external to the devices that affect their performance. Some of these parameters are at the origin of the internal degradation or are strongly related to this process. We can identify several external degradation mechanisms.

4.3.2.3.1. Material preparation parameters

These are parameters of the device manufacturing process. The deposition of the different layers of the device can be performed by various methods, and the adjustment of the parameters of each method used can affect the device and its operation. The choice and preparation of the substrate have a significant impact on the electrical behavior of the device because the substrate surface can cause short circuits if it presents excessive roughness. Furthermore, the substrate should be carefully cleaned to remove impurities that promote contamination of the deposited films. For vacuum evaporation of organic materials, films deposited under residual pressure may be contaminated by oxidation due to the presence of oxygen molecules or water molecules. For the deposition of polymers in solution, any contact with air or the water of the solution contaminates the films and annealing beyond the boiling temperature of water is recommended to remove water marks. In multilayer devices, the deposition technique must also take into account the possible damage created by the layer deposited on the previously deposited layer. If both layers are deposited by polymer solutions, the solvent used for the top layer must not dissolve the polymer from the bottom layer. If the top layer is deposited by vacuum evaporation, the deposition temperature should be adjusted so as not to enhance chemical reactions between the materials (Nguyen *et al.* 2006).

4.3.2.3.2. Degradation by impurities

Impurities are incorporated into devices during material synthesis or during device shaping. They act as non-radiative recombination centers and affect the processes and performance of the component. Impurities are frequently introduced into materials (active, dopant and metal) during their preparation and deposition. To reduce their concentration, purification methods should be used and purified materials enable the effects of impurities due to synthesis in device performance to be minimized. Other sources of impurities, such as residual air or water from the external medium, may also contaminate materials during shaping of the device films. The effect of impurities on the operation of devices is usually observed by a decrease in performance and lifetime.

4.3.2.3.3. Degradation by temperature

The glass-transition temperature, T_G, of organic materials is relatively low and the operating temperature range of organic devices is estimated on average between 50 and 100°C (Aziz *et al.* 2002). This temperature range is suitable for the use of devices exposed outdoors to the sun, that is, to organic solar cells. The rise in the temperature of the device during operation may come from the temperature of the external medium in which it is located, but may also be due to the Joule effect by dissipation of the heat released. Assuming that thermal processes are governed by the Arrhenius law, the effect of temperature on degradation is more pronounced when the temperature is high. For a thermally activated process with energy of 50 kJ/mol, the reaction rate will be doubled for a temperature increase of 10 K (Scholz *et al.* 2015). Since the Joule effect is produced by the current in the device, degradation (and therefore the lifetime) is a function of both temperature and device current intensity (Ishii and Taga 2002).

A *critical temperature*, T_C, exists, beyond which the device's operation is degraded with a sharp drop in performance. This temperature corresponds to the lowest temperature T_G of the materials of the electronic device (Tokito *et al.* 1996). When the temperature of the component is higher than the temperature T_G of an organic material of the device, the device is irreversibly degraded. It should be noted that some materials can undergo temperature-related transformations even when they are kept below T_G.

The first transformation process is the *crystallization* of the material, which changes the electrical and optical characteristics and thus the performance of the device. In general, an amorphous organic material tends to crystallize when stored in the dark if its temperature T_G is low. The same is true when the material is in contact with the electrodes, because in most cases the interface promotes crystallization of the organic material. The crystallization process is not often observed in materials of high temperature T_G that are more stable, demonstrating the benefit of using such material to extend the lifetime of organic devices.

The second transformation process is the *morphological evolution* of the surface of organic films. When an organic film is in contact with a substrate or an organic layer, the interaction between molecules in the two different media creates interfacial instability. The organic material is originally in metastable state and tends to stabilize over time. The surface of the film is gradually transformed under the effect of the temperature and its morphology gradually changes and stabilizes. In BHJ structures, the donor and acceptor materials are intimately mixed to obtain a large contact surface and thus increase charge dissociation and consequently the conversion efficiency of solar cells. However, the contact interfaces are unstable and the morphological evolution under the effect of temperature leads, first, to the gradual separation of the materials and then to the grouping of the molecules into clusters in the organic matrix. This process is called *phase separation* and is one of the causes of degradation leading to loss of efficiency in organic solar cells (Hoppe *et al.* 2004). The surface morphology of the active film of the BHJ structure is particularly important because it is in contact with a layer of different material and the interface so formed may constitute an additional source of phase separation (Nguyen *et al.* 2014). Morphological transformation is a slow process when the device operates under moderate temperature conditions. It can be accelerated during annealing at high temperature and can cause degradation by delamination of the electrode layers (Ahn *et al.* 2005). The annealing process applied to the devices normally improves their stability and lifetime, provided that the parameters (sequences, duration, temperature and pressure) are well-defined beforehand. Annealing, performed under the correct conditions, removes residual contaminants (oxygen and water) that are desorbed from the substrate or material and reduces defects in the layers.

4.3.3. *Degradation of OFETs*

The degradation of OFETs is manifested by the shift of the transfer characteristics over time during the transistor's operation, causing the threshold voltage to shift over time. This process impacts on the operation of the transistor V_T and can reduce its efficiency. For example, for an OFET used to control a pixel in an OLED screen, moving the transfer characteristic may cause the transistor quiescent point to switch and therefore the pixel to turn off. In an AMOLED active-matrix screen, each pixel is controlled by two transistors: T_1 to activate the pixel and T_2 to switch it on. A variation in the threshold voltage of 0.1 V can cause a 20% variation in the luminance of the pixel that depends on the drain current I_{DS} (In and Kwon 2009). OFET degradation measurements consist of continuously applying voltages V_{DS} and V_{GS} to the transistor and recording the transfer and/or output characteristics as a function of time (*voltage stress*). The changes to the characteristics enable the degradation processes in the transistor to be studied. In addition to the movement of the threshold voltage V_T with the operating time, other changes can also be observed, such as the increase in the subthreshold slope, the decrease in mobility of charge carriers and the increase in hysteresis.

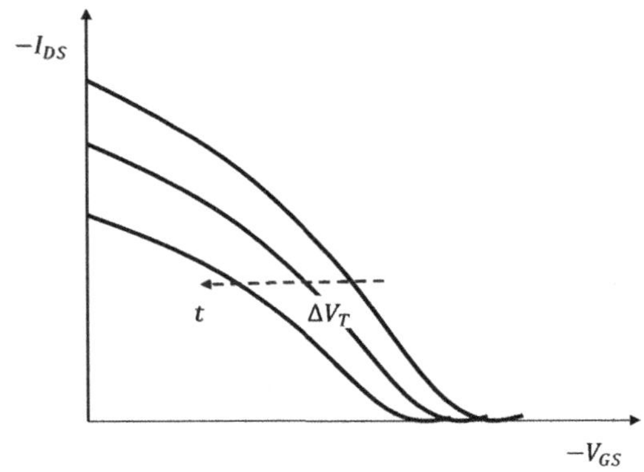

Figure 4.2. *Shift of the threshold voltage as a function of time due to the voltage stress effect*

4.3.3.1. *Role of intrinsic factors*

In a biased P-type OFET with a voltage $V_{GS}<0$, the threshold voltage, V_T, moves towards negative values during degradation. In other words, the shift of V_T occurs in the same direction as V_{GS}, that is, it becomes more negative as the transistor degrades. Nevertheless, the degradation is reversible and the characteristics return to their initial values when the transistor is left at rest for a sufficient period of time. When the transistor has been operating for a long time (several hundred hours), the voltage V_T may become saturated and the degradation becomes irreversible. At the origin of the degradation process, the trapping of charge carriers plays an essential role. Indeed, the current intensity, I_{DS}, decreases over time when a number of free carriers are trapped and no longer participate in transport. To obtain the initial intensity value of I_{DS}, voltage V_{GS} should therefore be increased, becoming increasingly negative as the degradation increases (Sirringhaus 2009). The shift, $\Delta V_T(t)$, of the voltage V_T is proportional to the concentration of the traps and is given by a *stretched exponential decay* function:

$$\Delta V_T(t) = [V_{T\infty} - V_{T0}]\left[1 - exp\left(-\frac{t}{\tau}\right)^{\beta}\right] \quad [4.1]$$

In this expression, $V_{T\infty}$ and V_{T0} are the boundary values of V_T at the moments t_∞ and t_0, τ is the relaxation time and β is the stretching exponent. For very long transistor operating times, the $\Delta V_T(t)$ curve tends to a limit corresponding to the saturation of the threshold voltage, V_T (Mathijssen *et al.* 2007).

Carriers captured by shallow traps are released quickly after bias is removed, which explains the recovery of the initial characteristics observed when the transistor is placed at rest. Carriers caught by deep traps would be responsible for irreversible degradation of the device.

Determining the traps in an OFET is therefore important, and their localization is possible in several regions of the device: in the organic SC, in the insulating film or at the SC/insulator interface.

4.3.3.2. *Role of extrinsic factors*

The extrinsic degradation factors of OFETs are similar to those found in other organic devices. The influence of the external medium on degradation is studied by experimentation with temperature stress (Kehrer *et al.* 2012), exposure to the humid atmosphere (Zilker *et al.* 2001) and illumination (Taylor *et al.* 2006), using the same protocol as that for degradation by voltage stress. Several hypotheses have been proposed to explain the OFET degradation process. They are based around the formation of traps in different parts of the device according to the stress conditions used. When the transistor is exposed to ambient air, the oxygen or water molecules diffuse into the organic SC and create electron traps that induce P-type doping and cause a shift in the threshold voltage. Similarly, molecules of impurities from solvents or ambient air can also diffuse into the organic SC and form trap centers as before, to degrade the transistor. Contamination of the insulator surface is very often invoked to identify the origin of the traps and the SC/insulator contact has been particularly widely studied. In particular, in a coplanar-structure transistor, the contact surface is small and near the source contact and the injected current intensity is high and causes rapid degradation. In contrast, in a staggered-structure transistor, the current intensity is lower due to the trapping of the carriers in the channel and the degradation by the contacts is less significant (Richards and Sirringhaus 2008).

Degradation processes involving the creation of trap centers in the organic transistor are generally put forward to explain the experiment results. However, analysis of the parameters of traps obtained in the studies leads to physical inconsistencies in several cases (Bobbert *et al.* 2012). Certain recent developments limit the role of traps, that is, they are only responsible for degradation in certain conditions and in certain materials. These developments focus on other mechanisms that take place at the contacts or in the insulator that highlight the transport of charge carriers in materials. In particular, the dispersive transport of holes in the insulator could be at the origin of the $\Delta V_T(t)$ curve observed experimentally.

In view of these analyses, different approaches and solutions have been proposed to increase the stability of OFETs, such as the use of encapsulations, passivation layers, new insulator materials and thermal treatments. These techniques are not only applicable to OFETs but to all organic electronic devices with varying degrees of success. As in other organic components, the stability of OFETs largely depends on the quality of the materials used and the best organic SCs remain to be found.

4.3.4. *Measuring the lifetime of devices*

The processes of degradation of materials, especially of organic devices, are complex, and degradation measurements should be made under identical conditions applying similar protocols to be able to compare the results obtained from materials and structures. In practice, it is difficult to conduct strictly identical experiments in different laboratories on the same materials or devices, and therefore the comparison of the lifetimes obtained by different teams remains quite unreliable.

Standards for measuring the lifetime of organic solar cells have recently been established, enabling streamlining of protocols and making result comparisons more credible (Reese *et al.* 2011). These standards specify the instructions and conditions for performing stability tests on solar cells. The recommended protocols are comprehensive and cover the different media (outdoors, in the dark or in the laboratory with simulated climatic conditions) in which the tests are carried out.

4.3.4.1. *Determining the lifetime of devices*

4.3.4.1.1. Different lifetime values

The degradation of a device results in a gradual decrease in the performance characteristic parameter over time. To determine the lifetime, T, the parameter defined as a function of time is measured by keeping all variables that can influence the measurement constant. For example, for OLEDs, we will plot the $L(t)$ curve representing luminance as a function of time. *Lifetime* is generally defined as the operating time for which the luminance, L_0, at the initial time has decreased by half, or 50% of its value, at a constant current intensity. In other words, it is the time corresponding to the value 0.50 of the $n = \frac{L(t)}{L_0}$ ratio of the luminance, $L(t)$, of the diode at time t to the luminance, L_0, at the initial time. This duration is denoted as T_{50}. Other definitions of lifetime have been proposed by adopting a ratio, n, above 0.50. The values chosen are then 0.70 or 0.80, and the corresponding lifetimes are denoted as T_{70} and T_{80} (see Figure 9.3a).

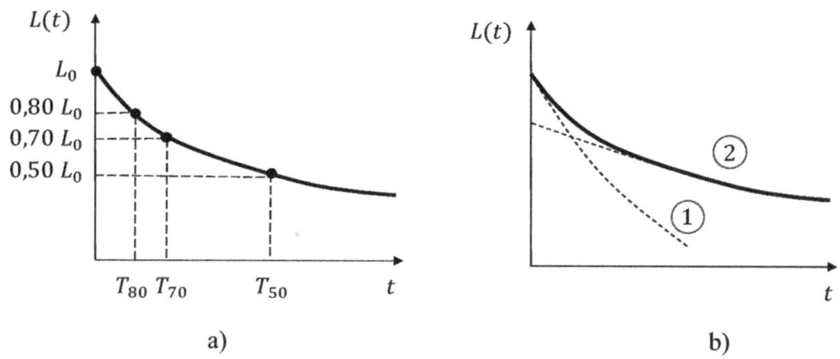

Figure 4.3. *Lifetime of an organic light-emitting device: a) different definitions of the lifetime of an OLED; b) the components of the $L(t)$ curve of an OLED*

The degradation curve representing the characteristic function of the device usually shows two phases of decay (see Figure 4.3b). Each phase can be described by an exponential function and, for an OLED, the evolution of luminance as a function of time can be expressed by:

$$\frac{L(t)}{L_0} = Ae^{-k_1 t} + Be^{-k_2 t} \quad [4.2]$$

where A and B are constants, and k_1 and k_2 are decay rates of degradation phases.

The first phase of decay is rapid with a sharp decrease in luminance and corresponds to an extrinsic degradation process, associated with the interfaces of the device. The second phase is slower, with a moderate decrease in luminance, and corresponds to an intrinsic degradation process, associated with losses due to the volume of the active layers (recombination centers and traps) (Turak 2013). This description remains schematic and does not sufficiently reflect the complexity of degradation processes, which are often underestimated when extrapolating the curves of equation [4.2] to determine the lifetime of the device.

Another approach to studying lifetime is the use of *stretched exponential decay* (SED), which is the complementary cumulative Weibull distribution function. This approach, based solely on experimental results, uses two parameters without physical significance: τ is the decay constant and β is the stretching factor or exponent. For an OLED, the $\frac{L(t)}{L_0}$ ratio is written as:

$$\frac{L(t)}{L_0} = exp\left[-\left(\frac{t}{\tau}\right)^{\beta}\right] \qquad [4.3]$$

The method enables the decay curves of the devices to be adjusted and their lifetime to be determined with greater accuracy than the simple exponential function method (Turak 2013).

4.3.4.1.2. Accelerated degradation test

The lifetime of marketable devices varies according to the application envisaged within a wide range of 10^4 to 10^7 hours. It is obvious that real-time testing is not always feasible and accelerated processes are used to be able to artificially determine component lifetime. We will also talk of *accelerated artificial aging* tests.

For an OLED, the initial luminance, L_0, and lifetime, T_{50}, are linked by an empirical relation (Féry *et al.* 2005):

$$L_0^n \times T_{50} = C \qquad [4.4]$$

where C is a constant and n is the acceleration coefficient, which depends on the nature of the material and the architecture of the device.

According to this relation, if the current, and therefore the luminance, is increased, the lifetime of the OLED will decrease. Similarly, the measurement of a diode's operating time can be extrapolated to determine its lifetime. For solar cells, the parameter used is power conversion efficiency, and the accelerated measurement principle is the same. The short-circuit current density J_{CC} will also be used to determine the decay constants and the open-circuit voltage V_{OC} is generally stable during cell operation. The choice of test parameters (the spectrum of the light used and the intensity, temperature and atmospheric parameters of the medium) may influence the degradation measurement and explains the divergence in the lifetime determination results. The use of accelerated degradation tests cannot therefore be considered as a standardized process for studying the behavior of materials and devices with respect to degradation processes (Bobbert *et al.* 2012).

4.3.4.2. *Improving the lifetime of devices*

Understanding the processes of degradation of materials and devices makes it possible to propose solutions to mitigate the effects of the different factors that cause or promote degradation reactions. It should be recognized, however, that there is no universal solution to eliminate all causes of degradation and ensure long component lifetime. Nevertheless, a number of global solutions can be applied to solve component degradation problems and other specific solutions are effective for particular materials and structures.

To mitigate the degradation process by diffusion of oxygen and water in materials, the encapsulation technique (by means of a glass slide or thin encapsulation layers) is used to delay the diffusion of oxygen or water molecules. The technique is effective, but the encapsulation layer is bulky and rigid, therefore inappropriate for flexible components.

To prevent chemical reactions at the ITO electrode, the oxide layer can be treated to stabilize the contact surface. Acid, ozone or UV treatments on the oxide surface are often effective but may alter its work function and influence the performance of the device. Direct contact between ITO and organic films can be avoided by inserting a buffer or intermediate layer that separates the two surfaces. This layer, generally very thin (~ 5 nm), is a metal compound and often an oxide (ZnO, TiO_2, Al_2O_3, LiF, CaF_2, etc.). It enables the lifetime and performance of the device to be improved. The buffer layer can also be deposited between the metal electrode and the organic film. It also acts as a mechanical reinforcement for the adhesion of the metal electrode to the structure and attenuates the delamination process.

Replacing ITO with another conducting transparent material (PEDOT:PSS or other metal oxides) can also be considered, but the results are not conclusive at present.

Many other solutions exist that are applied to specific materials and structures to improve component stability. For example, doping of active and transport layers prevents the accumulation of charge carriers that degrades materials. The use of inverted structures in place of standard structures solves the problem of degradation of the interfaces between the electrode of reactive metal (Ca, Ba) and the organic film. In general, the quality of the materials used in the device plays an important role in the durability of the component. Good electrical, optical, thermal and mechanical properties are required to prevent premature material degradation during operation, which would result in failure of the device.

The standard lifetime of OLEDs and OSCs for marketing is estimated at 10,000 hours. For OLEDs, this time has been set to T_{50} for display screens with a luminance of 100 cd/m^2. For lighting, it corresponds to T_{70}, with a luminance of 1,000 cd/m^2. Indeed, it is found that the use of the T_{50} parameter for light fixtures may produce non-uniform lighting due to premature aging of certain groups of pixels or of certain diodes in a panel. For domestic applications, the daily and average usage time of light fixtures is estimated at 5 hours, while for industrial use, it is 10 hours per day (Tyan 2011). For a lifetime of 15 years, WOLEDs should operate for 27,000 or 54,000 hours. These values have already been reached with the T_{50} parameter. Improvements in performance (materials, architecture and efficiency) will be necessary to achieve the performance required for these devices.

For OSCs, a lifetime of more than 10,000 hours at T_{80} has been obtained on cells tested in the open air (Gevorgyan *et al.* 2013). For a daily operating time of 12 hours, this lifetime represents two to three years, which is very short compared to the lifetime of current inorganic solar cells (~20 years). Several projections use an estimate of 1,000 hours of annual operation, and for 10,000 hours (Krebs and Spanggaard 2005), a lifetime of 10 years is estimated. Taking into account the results of tests carried out by different teams, it would be more reasonable to estimate a duration of around five years (Brabec *et al.* 2005) as a basis for cells operating outdoors.

Despite progress made in understanding the degradation processes of organic solar cells, and despite treatments and solutions to improve the performance of these cells, their limited lifetime currently remains a problem that needs to be solved to enable large-scale marketing. The use of cells indoors is another opportunity to exploit energy conversion and offers real marketing opportunities. Indeed, the difference between outdoor and indoor solar cell usage is the light intensity and spectrum of the light source. Outdoors, the cells operate under solar irradiation at AM1.5G, while indoors, they operate mainly under the light of light fixtures with intensities less than 1 $mW.cm^{-2}$, that is, below 1% of the solar intensity. In addition, the lamp emission spectrum presents bands of high intensity in the range of visible wavelengths. Therefore, the power delivered by cells indoors would be lower than that delivered by cells operating outdoors. It would, however, be enough to power portable devices in the *Internet of Things* (IoT) system, which is the interconnection of objects connected to the Internet (computers, wireless sensors, mobile phones and tablets) (Lee *et al.* 2016). In addition, the structure of the cells could be simpler to absorb radiation from lamps whose characteristics are virtually constant and independent of weather and climatic conditions of the medium. The cells used indoors are manufactured using the same materials and structures as conventional cells, but their efficiency is higher. The improvement in performance is explained by the high value of the parallel or *shunt* resistance, R_P (Steim *et al.* 2011). Cells based on a mixture of [6,6]-phenyl C71 butyric acid methyl ester ($PC_{71}BM$) and poly[N-9'-heptadecanyl-2,7-carbazole-alt-5,5-(4,7-di-2-thienyl-2',1',3'-benzo-thiadiazole] (PCDTBT) provide a 16.6% conversion efficiency indoors. A solar cell with this efficiency and surface area $10 \times 10 = 100$ cm^2, exposed to a light source of intensity 1 $mW.cm^{-2}$, would provide a power of 16 mW, which is sufficient to supply low-power devices.

4.4. Organic device production cost and marketing

The selling price of an industrial product includes the cost price, taxes and margins. The cost price consists of direct expenses, that is, expenses due to the manufacturing process of the product, and indirect expenses, that is, overhead costs due to the business operations.

Direct expenses refer to both raw material and labor charges. For electronic devices, the materials used are mainly polymers or small molecules whose preparation requires the use of organic solvents. For substrates, TCOs are deposited on glass or plastic slides. The price of raw materials is a relatively high part of the production cost. Consider as an example the manufacturing of a solar cell based on P3HT:PCBM in the laboratory. The price of the P3HT polymer is ~500 €/g, and the price of the fullerene is ~8,000 €/g. To make a solution (1:1) of concentration 10 mg/ml, the cost of the solution is 65 €/ml and a volume of 100 µl is used for deposition by spin-coating, that is, there is a cost of €6.5 for the active layer deposition. Added to this sum are the costs of transport films and electrodes and the cost of the cell will be of the order of €15 to €20. This is only a rough estimate of the cost of the materials used to produce a P3HT:PCBM cell at laboratory level. For industrial production, the calculation of the cost of raw materials is more elaborate and different because of the deposition technique and the quantity of cells manufactured.

Although devices such as OLEDs, OSCs and OFETs are new, inorganic components assuring similar functions have already been marketed. Organic devices arriving on the market are not novel but are competitive with existing ones. As such, they have the disadvantage of being newly competitive products, i.e. they must demonstrate their efficiency and advantage over existing products in order to be adopted by consumers.

We will examine the production conditions of each component category and compare them with those of the corresponding inorganic devices.

4.4.1. *Production of OLEDs*

In the field of display, OLED screens have moved from small screens of portable devices to television screens in competition with LCD screens. In addition, rollable screens have recently been put on the market, demonstrating the capabilities of OLEDs in electronic applications.

In the field of lighting, conventional technologies such as incandescent lamps and fluorescent lamps have been progressively replaced by LED technology, which uses III–V or II–VI compounds for the emitting material. LED lamps present undeniable qualities in terms of lifetime (20,000–50,000 hours) and efficacy (100 lm/W). The problems with LED use concern the temperature control of diodes by heat dissipation via the substrate, which has to be a good thermal conductor. The crystallographic structure of the substrate should be compatible with that of the emitting SC to avoid the formation of defects in the SC during its growth. The materials used are sapphire (inexpensive) or silicon carbide (expensive). The lack of temperature stability leads to a drift in the chromatic coordinates, especially with RGB technology. To improve stability and achieve good LED performance, high-quality materials should be used, thus increasing the manufacturing cost.

Assuming comparable lamp installation and maintenance costs for LEDs and OLEDs, the comparison of light fixtures should take into account the *purchase cost* and the *cost of electricity consumption* (or use) for each type of lamp. Since the characteristics and performance of the lamps are different, production cost is expressed in dollars per lumen ($/lm), that is, the amount to be spent to obtain a luminous flux of one lumen. In the long term, the lifetime of the lamps should also be taken into account as it impacts the cost of purchasing replacement lamps in the event that a lamp fails.

For inorganic LEDs, the production cost is estimated at ~0.01 $/lm (So *et al.* 2008) and their lifetime is estimated at ~30,000 hours, with an efficacy of ~120 lm/W. As discussed above, the characteristics of WOLEDs are quite similar to those of LEDs, with the exception of the lifetime (~15,000 hours), which remains to be improved. For an OLED module of surface $S = 9\ cm^2$, efficacy = 64 lm/W and luminance = 4,000 cd/m^2, the price should be less than $10 to be competitive on the market (Steim *et al.* 2011). The production cost of OLEDs for lighting can be significantly reduced if we consider the possibilities of mass production by printing (by roll-to-roll processing or ink-jet printing), even though the quality of the printed matter is currently not perfect. Indeed, for a lamp, high resolution is not required as it is for display screens. On the other hand, a degraded pixel does not affect the overall luminance of the emitting surface. Other possible solutions to reduce the production cost would be to use substrates with a TCO less costly than ITO. The arguments in favor of OLEDs are, on the one hand, the growing need to develop efficient and economical lighting sources (on average, 20–30% of electrical energy is used for lighting worldwide) and, on the other hand, the ability to design lamps of various shapes and sizes that no other technology can achieve.

4.4.2. *Production of OSCs*

As with OLEDs, the issue of OSC production is linked to the solar cells on the market, which are mainly silicon-based cells. These cells have a high conversion efficiency (~25%) and lifetime (~20 years), as well as a high manufacturing cost. To compare the costs of different technologies, performance (efficiency, lifetime and cost) is expressed as a function of a parameter called *Watt-peak* (Wp). This is the maximum electrical power produced by a solar cell of surface of $S = 1\,m^2$ maintained at temperature $T = 25°C$. For polycrystalline silicon-based cells, the cost of the modules is ~$2.34\ \$/Wp$ (Kalowekamo and Baker 2009), assuming an efficiency of 15%. The performance of OSCs is very close to that of silicon cells, but their lifetime is still significantly lower than that of silicon cells. The estimated 10 years of operation for organic cells is rather optimistic and a lifetime of five years is more realistic. However, there are strong arguments in favor of significant development of organic cells in the near future. These arguments are based essentially on the possibility of reducing the production cost of these cells, which would enable them to compete with silicon cells. Indeed, Brabec has compared the production of solar cells with a total surface area of 88,000 m^2 by silicon technology and by organic film printing technology (Brabec 2004). The time required to produce silicon cells is one year, and the time required for organic cells is a maximum of 10 hours. The cost of manufacturing organic cells could be considerably lower than that of silicon cells. However, a comparative study of the cost of producing cells using different technologies showed that the cost of an organic cell varies between ~1.0 and 2.83 $/Wp; in other words, it is still very high (So *et al.* 2008). In this study, the efficiency of OSCs was 5% and their lifetime was five years. As we saw above, current cells present an efficiency greater than 5%. For single-component organic cells, it is better than 15% and that of tandems cells exceeds 17%. Perovskite solar cells have even higher efficiencies: more than 20% for single-component structures and more than 23% for tandem structures. These values are obtained for cells created in the laboratory and suggest that the cost of the modules should be lower than that in the study cited. More recently, a cost analysis of modules with different technologies placed the cost of organic cells at 0.27 $/Wp with the hypothesis of a lifetime of 5–10 years and a maximum efficiency of 7% (Gambhir *et al.* 2016). In the same projection study, the costs of modules obtained by other technologies (crystalline silicon, cadmium telluride or copper, indium and gallium selenide) varied in the range of 0.35–0.60 $/Wp. This result means that organic cells represent a promising solar power sector in the near future, provided they have a longer lifetime than today. On the other hand, it will be necessary to consider the outdoor operating conditions of the cells to determine their effective lifetime. The use of encapsulation layers for effective protection of cell elements seems necessary, and the study of materials intended for encapsulation should be developed.

Another argument in favor of the future of organic cells is the possibility of producing large numbers of components with the printing technique. Conventional deposition techniques are well-known and mastered and allow for the production of high-resolution, uniform and identical cells, while there are divergences in results on the resolutions and performance of cells achieved with the printing technique. Recently, the progress made in this technique has made it possible to control the deposition of films with sufficient precision to obtain structures comparable with those obtained by evaporation or by the spin-coating method (Søndergaard *et al.* 2012). The printing technique does not appear to affect the solar cell lifetime. An advantage of this technique is that the encapsulation layer can be deposited directly on the cell after the last electrode has been deposited. The only difficulty lies in the deposition of the transparent conductive electrode, which could be replaced by other less efficient materials.

According to *Swanson's law* (Swanson 2006), when the production of cells doubles, their prices tend to decrease by 20%. While several studies predict a cost of less than 1 $/Wp for silicon cells, the goal of reaching the threshold of 0.27 $/Wp for organic cells with mass production and improved lifetime looks achievable in the near future.

4.4.3. *Production of OFETs*

For comparison, the best OFETs offer mobility performance and an I_{on}/I_{off} ratio equivalent to those of amorphous silicon transistors. On the other hand, their operating voltage remains high (20–50 V), at least for planar structures. The lifetime of OFETs has not been systematically studied. It should be recalled that there are two organic transistor degradation processes: voltage stress degradation and degradation by exposure to air or to the humid atmosphere. The second process is irreversible, and the transistor should be protected from the diffusion of water or oxygen molecules to prolong its lifetime. The use of passivation layers limits this diffusion and helps to enhance the stability of OFETs regardless of their structure. The PVA/acryl passivation bilayer used in a bottom gate/top contact (BG/TC) structure with a pentacene-based transistor enables the component lifetime T_{50} to be extended to 11,000 hours (Han *et al.* 2006). Similarly, a TG/TC structure using pentacene as the organic SC and a bilayer of Al_2O_3/CYTOP (fluoropolymer) as the dielectric shows stability for more than 5,000 hours without significant change in carrier mobility and threshold voltage (Hwang *et al.* 2011). These studies show that, with effective protection, the stability of OFETs is sufficient to ensure continuous and long-lasting operation.

Like other devices, OFETs are likely to be mass produced with printing techniques to reduce the production cost. However, the realization of transistors presents certain problems, including the resolution and alignment of printed patterns. Indeed, when the resolution and accuracy of the transistor elements are low, the electrical characteristics may vary from component to component and it is not possible to ensure the uniformity of the transistors in a circuit using a large number of components. Techniques for preparation of the substrate, on which the source and drain electrodes are deposited, have been proposed to increase the print resolution to 5 to 10 μm for the length of the channel of the OFET (Schmidt *et al.* 2010). The results show that the roll-to-roll printing technique can produce circuits with high integration density and acceptable resolution.

One of the economically significant applications of OFETs is the manufacture of RFID identification tags that allow data to be stored and retrieved remotely. Each tag contains several thousand transistors, and it will be of more interest to use OFETs than conventional silicon-based transistors to lower the production cost. The annual tag market is valued at more than $10 billion, and with the development and growth of the current market, it is likely that OFET technology will be called upon to meet demand. With silicon-based transistors, a passive tag costs less than $0.10 to produce. Producing the same tags at a lower cost with OFETs will require limiting element resolution, even if the performance and uniformity of transistor characteristics are reduced.

4.5. Synthesis on Brabec's criteria

Having examined the different aspects of the three main electronic devices, OLEDs, OSCs and OFETs, according to the criteria set out by Brabec, it is clear that these components will continue to increasingly occupy the electronics market sectors. OLEDs are already widely used for portable device display screens and television screens. They are also produced for household lamps and could potentially constitute light sources for buildings and public places in the near future. The lighting market looks very promising for the development of this technology. Despite progress made in the fields of materials and cell architecture, OSCs have not yet reached the standard sought for marketing because of the lifetime limits, which remain insufficient. Increasing cell longevity and operational stability is essential to be able to compete with other technologies, especially silicon cells, which can be operational for more than 20 years. Despite these limitations, OSCs have been successfully marketed for portable or low-power devices. For OFETs, applications in AMOLED matrices have demonstrated the potential of organic transistors in the electronics market. The use of these components in the manufacture of RFID tags is also promising. For all of these components, the potential for economic development lies in mass-production techniques, which are still to be improved to ensure uniformity and reliability of the components. While device production has been mastered, there is no doubt that the market for organic electronic components is becoming one of the most attractive for high-tech products in the future.

Many other organic electronic devices have been produced and studied, which we have not examined here. Some of these devices are of great interest and look promising (photocatalysts, batteries, sensors, etc.) and will be the subject of future work.

4.6. Environmental dimension

Organic electronics is a relatively new technology, and the devices we examined in Chapters 1, 2 and 3 have not yet been manufactured and marketed at the same level as those derived from silicon technology. As a result, while the impact on the environment of the use of conventional electronic components and their disposal after use is fairly well-known, the same effects from organic components are not yet widely studied and are even unknown for certain aspects (Nguyen and Yang 2018).

Consider as an example the use of the materials used in mobile phones. These devices have a service life of about three years and are produced in large quantities worldwide. Once they become obsolete, they are withdrawn from the market and disposed of in an organized or unorganized manner. Waste from these devices contains toxic substances and, if dispersed in nature, there is a risk of contamination of surrounding environments, soil, air and water, which in turn contaminate living beings and biota. If we look at the problem in terms of the manufacturing of these devices, we note that it also raises the question of the energy spent to produce the electronic components necessary for the realization of the device and, indirectly, the consumption of energy resources and the waste generated from it. In other words, the acquisition and use of electronic devices (organic or otherwise) have consequences on the environment, and the effects produced should be considered, studied and analyzed in order to respect and preserve the nature and environment in which we live.

It is clear that the issue of the environment needs to be dealt with much more comprehensively in accordance with the resolutions of international treaties on climate change and, in particular, *sustainable development*. This terminology refers at the same time to growth in long-term economic activity, a response to the basic needs of the society in which people live, and respect for the environment. In this spirit, the development of electronic devices should meet the ecological dimension criterion, namely the quest for harmony between humans and nature.

In this section, we discuss the environmental aspects of organic electronic component technology to understand the issues relating to life cycles for products using this technology, on the one hand, and the impact of these products on the environment, on the other hand.

4.6.1. *Life-cycle assessment*

This analysis makes it possible to assess the environmental impact of a system or product during its lifetime, taking into account all the elements and facts enabling its realization, up to the end treatment of used components at end-of-life. A product has two distinct life periods:

– the cradle-to-gate period, from production to factory gate;

– the gate-to-grave period, from factory gate to disposal.

For each period, the stages are defined and the processes with effects that influence the evolution of the environment are studied for each stage.

For example, for the study of OSC modules, the cradle-to-gate period consists, schematically, of the following stages:

– acquisition of materials: polymers, metals, substrates, solvents, etc.;

– synthesis (where applicable) of organic or composite materials;

– cell manufacturing: deposition of layers, treatments, etc.;

– module manufacturing: cell welding, package mounting, etc.;

– performance tests: stability, efficiency, lifetime, etc.

The gate-to-grave period consists of the following stages:

– use of cells and modules;

– end-of-life management: reuse, recycling, destruction, etc.

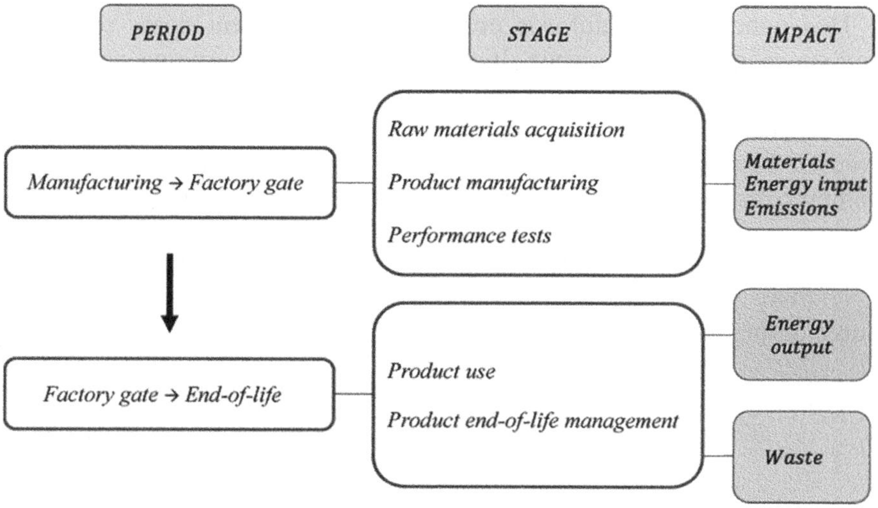

Figure 4.4. *Diagram of the life-cycle assessment of a solar cell*

For each stage of the study, the impact of realization (materials used, energy consumed and waste) on the environment can be evaluated by determining measurable categorical parameters, such as emission rates of CO_2 gas (climate change), CFC gas (ozone layer), acidification, etc., for a produced power of 1 Wp by a cell with a given lifetime. The production of toxic substances such as metals and chemical waste is also evaluated, and the sresults are taken into account in the product's life-cycle assessment.

The comparison of different types of solar cells with respect to the environmental impact makes it possible to decide on the choice of technology to be favored in accordance with sustainable development criteria.

In general, the analysis carried out enables identification of problems or points that can or should be resolved to improve the product. It also helps to contribute to the development of other new products. There are many factors or indicators used in the analysis of a product's life cycle. Here we discuss two that are frequently used for the study of electronic components: the *levelized cost of energy* and the *energy payback time*.

4.6.2. *Levelized cost of energy*

This concept is useful for evaluating the different costs required to produce energy from a source. It measures the cost of energy produced by the source over its entire lifetime and therefore, in comparison with other sources, the efficiency and cost-effectiveness of the source concerned. The *levelized cost of energy* (LCOE) includes the cost of initial investment and running costs which include operating, consumable and maintenance costs.

This indicator is obtained by dividing the levelized sum of the expenditure by the levelized sum of the amounts of energy produced by the source during its lifetime (n years):

$$LCOE = \frac{Total\ cost\ of\ energy\ source\ (n\ years)}{Total\ amount\ of\ energy\ produced\ by\ the\ source\ (n\ years)} \qquad [4.5]$$

It represents the amount to be spent to produce energy of one kilowatt hour and is expressed in units of $/kWh.

4.6.3. *Energy payback time*

The *energy payback time* (EPBT) of an energy generation system is, by definition, the time (in years) that the system needs to operate to generate an amount of energy equal to that consumed to build it. The evaluation of this indicator depends on the materials used in the system, the energy efficiency (power conversion efficiency for solar cells) and the climatic conditions of the system's place of operation.

4.6.4. *Life cycle of organic solar cells*

Among the organic electronic devices studied, OSCs generally operate under climatic conditions that frequently expose them to higher risk of degradation than OLEDs or OFETs. Indeed, the cells are designed to transform light energy into electricity in an economical and ecological way. They are placed outdoors and under sunlight in all weather and can be subjected to the effects of extreme heat, extreme cold or rain. Unlike OLEDs and OFETs, mechanical and thermal stresses in OSCs can cause damage to the material, cell, module and protections (encapsulation and housings). For this reason, the OSC life-cycle study is considered to be representative for organic electronic devices.

Several studies have focused on the life cycle of OSCs in comparison with that of cells derived from traditional technologies (silicon, CdTe or CIGS). The following example has been drawn up according to the hypothesis of 25-year operating cells with a power of 1 Wp. The environmental impact of production of 1 Wp by an organic cell with an efficiency of 5% has been compared with that of a silicon cell with an efficiency of 13.2% (Roes *et al.* 2009). The organic cell is assumed to have the structure ITO/(PEDOT:PSS)/(P3HT:PCBM)/LiF/Al, deposited on a glass substrate. Plastic substrates (poly(ethylene terephthalate), or PET) are also used and the structure of the cell studied is ITO/(PEDOT:PSS)/(P3HT:PCBM)/Al; a double encapsulation in composite epoxy/SiO_2 is added to protect the cell. The costs of materials used in the cell are evaluated from the dimensions of each layer and the cost of each material purchased or manufactured. The cost of energy spent for substrate preparation is also evaluated, followed by the cost of the housings and their installation.

If one compares the life cycle of organic cells (on glass or plastic substrate) with that of multi-crystalline silicon cells, the results show a much less significant environmental impact for organic cells. Taking into account the lifetime of silicon cells (25 years) and organic cells (5 years), more organic cells will need to be used to produce the same amount of energy produced by a silicon cell for a duration of 25 years. The result indicates that the consequences on the environment will be greater for organic cells.

With respect to cost, the largest share of organic cells comes from the assembly of the modules and the total cost decreases as conversion efficiency increases. For cells on glass substrate, the cost is 4.18 €/W_p for a conversion efficiency of 5% and 1.90 €/W_p for an efficiency of 11%.

This example shows that organic cells can only compete with conventional cells if their efficiency is higher and their lifetime is longer.

4.6.5. *Fate of released pollutants*

The effects on the environment induced by the use of an electronic component may be examined through the pollutants produced during and after its lifetime. These pollutants, harmful to surrounding environments, include the gases emitted (CO_2, CFC) and the materials (organic, inorganic and metal) used in the component. They can diffuse in the air, soil and sediment and contaminate water sources and are then consumed by living organisms. In this section, we examine the pollution effects of organic solar cells due to their conditions of use, which are more frequently conducive to premature degradation. It is evident that the pollution caused by other organic devices is as significant as that originating from solar cells, and their effects on the environment are also comparable.

4.6.5.1. *Degradation of cells and modules*

Under normal operating conditions, the solar modules are assembled in a housing that protects the cells from bad weather when used outdoors. The cells are encapsulated by a thin layer (a glass or plastic slide) impermeable to water and air. In principle, there is no gas emission or water ingress into the cells and housing during operation. However, it is not to be ruled out that mechanical incidents could occur, resulting in the destruction of modules or degradation of devices (aging, wind, storms or other events).

At the devices' end of life, they can be recycled, destroyed or dispersed in nature according to each country's environmental policy. Recycling is the best solution to minimize pollution from waste, but the organization of treatments is costly. For the other cases, it is necessary to study what becomes of the components and determine the significance of their impact on the environment.

4.6.5.2. *Emission of greenhouse gases*

Greenhouse gases (GHGs) are gases emitted into the Earth's atmosphere that absorb infrared radiation. Part of the heat stored by these gases is dissipated into space, and the remaining part produces a greenhouse gas effect that contributes to a rise in the Earth's temperature. This phenomenon is known as *global warming*. These gases are mainly water vapor, carbon dioxide (CO_2) and methane (CH_4). The origin of GHG emissions can be natural, originating from activities of the Earth (evaporation of seawater, volcanism and breathing of living organisms) or due to human activities (energy consumption and waste production).

To evaluate and compare the importance of gas emissions and their impact on the environment, a standard unit, called the CO_2 *equivalent* or *global warming potential* (GWP) is used. This is the mass of CO_2 gas that would produce an equivalent impact on the greenhouse gas effect or global warming over a defined period of time (generally 100 years). For CO_2, the GWP equivalent is 1 and, for example, CFC-12, or dichlorofluoromethane, has a GWP of 10,900 for the duration of 100 years.

For organic solar cells, GHG emissions over their lifetime are derived from electricity and (fossil) fuel consumption and module production. To compare the effects of the production and use of solar cell technologies, the CO_2 *emission factor* is defined as the ratio of the CO_2 equivalent total of a module to the amount of electricity generated over its entire lifetime. Thus, this factor indicates that the environmental impact of a solar cell depends on its lifetime and efficiency. In the hypothesis of organic cells with 1% efficiency and a two-year lifetime, the CO_2 emission factor is almost double that of crystalline silicon-based cells (Darling and You 2013). In the long term, that is, by the end of the 2020s, the trend will be reversed, with improvements in efficiency (~15%) and lifetime (~20 years). The CO_2 emission factor of organic cells would be 10 times lower than that of silicon cells. In other words, as long as the efficiency and lifetime of organic cells remain low, it is not desirable to consider their production, and it would be preferable to manufacture silicon cells to exploit solar energy, as the environmental impact of this technology is more favorable.

4.6.5.3. *Toxicity of polluting materials*

Materials resulting from the decomposition of electronic devices can be classified into two categories: inorganic materials and organic materials.

4.6.5.3.1. Inorganic pollutants

Inorganic materials can be toxic in various forms:

– nanomaterials that are used in composites or mixtures (P3HT:PCBM, for example) are still unknown from an environmental perspective, especially their fate in aquatic environments (Handy *et al.* 2008). They are easily transported by water from rivers and incorporated into living organisms. The effects of nanoparticles on the human body are not yet well known, and even titanium dioxide, which is widely used in components and authorized in foods as an additive, has recently been suspected to be carcinogenic. Other nanoparticles can be produced by the decomposition of metals serving as electrodes (ZnO, Ca, Ag) and then transported by water and ingested in food;

– metals constitute a potential source of toxicity for the environment. They are mainly found in the electrodes of components. First, ITO, which serves as a transparent electrode for OLEDs and OSCs, is unstable in the presence of water and decomposes by releasing indium and tin atoms. ITO particles cause inflammation and cancer in living organisms. Metals such as gold, silver or zinc can damage the cell structure of organisms until they are destroyed. The metal oxides used in components are also toxic to organisms. In particular, zinc oxide ZnO has been the subject of studies on the degradation of organic cells because it is used as a material for the nanostructured substrate in many cells. Oxide ZnO is recognized as responsible for inflammatory and cytotoxic effects in the human body. In the case of CH_3NH_3 PbI_3-based solar cells, the risk of lead diffusion in soil and water should be considered carefully as lead is toxic to the nervous system and immune defense (Hailegnaw *et al.* 2015). Lead concentration in perovskites is low, but solutions using other metals as alternatives to lead have been considered.

4.6.5.3.2. Organic pollutants

Organic materials are also harmful to the environment when dispersed in nature. The degradation of these materials is promoted by the combination of thermal, chemical, optical and moisture effects of the medium in which they are exposed.

The main organic materials to consider are the following:

– poly(ethylene terephthalate), or PET, is used as an encapsulation or substrate material and is potentially toxic to reproduction. This plastic is non-biodegradable, but may disintegrate into microparticles and be ingested by animals or fish and thus incorporated into the food chain. PET recycling techniques are well-established and can be used to minimize contamination;

– the fullerene PCBM and its derivatives, which are nanoparticles initially not soluble in water. The surface of these materials can become hydroxylated and make them soluble in water, and they can consequently be transported and contaminate the sources;

– the toxicity of conjugated polymers has not been widely studied, and the high chain-length prevents them from being absorbed by living cells. Material degradation can break polymer chains into small segments and the degraded products can become potentially toxic. The use of solvents to dissolve polymers also poses a serious problem for the environment. Common solvents for conjugated polymers are chloroform, chlorobenzene, hexane, toluene, xylene and other solvents chosen according to the polymer structure. For the printing technique, polymer ink is diluted in ethanol, butanol or isopropanol. On a practical level, solvents are volatile and evaporate at room temperature. Emissions of organic compounds into air, water or soil are dangerous because they are toxic products that are potentially carcinogenic. Solvent handling is regulated and recycling of used solvents is mandatory for laboratories.

4.6.6. *Mass production and environment*

Analyses of the environmental impact of solar cell technologies have shown that short-lifetime, low-efficiency organic cells do not meet industrial development criteria while protecting the environment and cannot replace cells produced by conventional technologies. While hoping and waiting for longer cell lifetimes and efficiency, the possible solution to making the use of these cells compatible with ecological goals would be to increase the amount of electricity generated by increasing the number of cells manufactured. This solution is based on the mass production of cells by printing techniques. The following advantages of using these techniques can be cited:

– thin layers can be printed on a variety of substrates: glass, plastic, sheet metal and paper;

– the process is straightforward and economical: thin layers are deposited directly on the substrate using a small amount of material (printing ink). A mask is not used to delineate the shape of the layers to be deposited;

– waste production can be expected to be low for the small amount of materials used.

From a technical point of view, the resolution of printed patterns is not comparable with that of the inorganic SC etching technology and, because of the defects in the printed image, there are risks of short-circuits or circuit breaks that would prevent the devices from functioning. On the other hand, the electrical contacts are printed with a colloidal ink, which is a solution containing metal particles (copper or silver). High-temperature annealing (~200°C) should be performed after the ink has dried to obtain good electrical conductivity. The substrate material needs to withstand the required thermal treatment without damage. The realization of the devices would require the use of clean rooms and equipment appropriate for mass production.

From an environmental point of view, several factors should be considered to evaluate the effects of printing technology. Unlike the technical and technological issues with printed electronics, there have been few in-depth studies focusing on the environmental assessment of this technology (Kunnari *et al.* 2009). First, mass production of organic devices means that a large amount of waste will be produced and will have to be treated. Second, printed electronics products generally have a short lifetime of a few years. This is due, on the one hand, to the technology that is not yet advanced enough to achieve it and, on the other hand, to mass production, which also targets commercial applications whose lifetime does not exceed 10,000 hours (Krebs and Norrman 2007). As a result, the number of used and disposed-of products can be expected to increase sharply as a result of the industrialization of printed electronics. During manufacture, emissions of solvent vapors are significant and should be controlled to avoid polluting the atmosphere of the surrounding environment. After the end of life of the devices, the problem of treating large quantities of waste arises. The following solutions may be considered (Keskinen and Valkama 2009):

– re-using the product with preservation of its function, to give it a second life. This solution does not correspond to the purpose for which printed products are typically used;

– recycling with material recovery to save natural resources and limit pollutant emissions. For printed products, the difficulty would lie in separating the different materials used in each layer;

– recovering energy by waste incineration to produce electrical or thermal energy. The risk of this solution is air pollution with the emission of potentially dangerous gases (carbon monoxide (CO, dioxin) or pollutants (CO_2, NO_2);

– dispersion in nature, which introduces pollutants or contaminants into the soil and alters its biological, physical and chemical properties. Polluted soil in turn contaminates water sources and then living organisms. For printed products, common pollutants that are potentially hazardous to health are metal or organic nanoparticles (fullerene), plastic particles (degraded substrates) and chemical waste.

In summary, although some studies have shown that organic devices (in this case, solar cells) obtained by printing technique on plastic substrates have a much lower environmental impact (from 80 to 95% lower) than equivalent devices manufactured by conventional techniques (Schmidt *et al.* 2010), there are uncertainties still to be elucidated as printing technology is evolving with the use of new materials and processes that could alter the indicators. For example, the application of *green chemistry* (sustainable or ecological) (Burke and Lipomi 2013) to the manufacture of new materials that can replace the materials typically used in the composition of films could significantly reduce the toxic substances released by waste. The principles of this chemistry aim to reduce or eliminate substances and processes harmful to the environment through new synthesis techniques, the use of new solvents and the use of new materials. In the field of solar cells, fullerene in the P3HT:PCBM mixture is known as a potentially toxic substance. In addition, the energy required to manufacture this material is several dozen times higher than that used to produce P3HT. To reduce the environmental impact of cells using the P3HT:PCBM mixture, it will be necessary to replace fullerene with other materials that require only a small amount of energy for production and are not harmful to health.

4.7. Prospects and developments

The first OLED screen-equipped digital cameras were produced by Sony in the early 2000s. The marketing of these screens proved a failure, and they were withdrawn from the market shortly after their introduction due to the instability of the diodes. A decade later, Sony launched the first OLED screen used in televisions of the same brand. These televisions were presented as having the thinnest screens on the market. After a short period on the market, the brand ceased production of these devices because tests revealed a much shorter lifetime than that announced. These two examples validate in part the criteria proposed by the Brabec triangle, namely the efficiency and lifetime of an organic device with a view to its marketing. To date, there has been no example of production being abandoned owing to the high cost of manufacturing the product. As we have mentioned, the price of any product marketed is initially higher than that of competing products already on the market. Nevertheless, as a new product with new and different features, it can attract and be purchased by consumers despite its price. Once the investment depreciation is sufficient and production becomes more controlled and substantial, the cost decreases and the product can be marketed on a larger scale.

Just before the start of the 2000s, a storm, described as the storm of the century, swept across parts of Western Europe and devastated entire regions. This was a phenomenon never seen before that many scientists understood to be a consequence of climate change, the cause of which was considered to be excessive greenhouse gas emissions. Until then, governments had discussed how to regulate GHG emissions (Kyoto Protocol in 1997) but a strict control over compliance with the agreement was never established. Local and national movements in favor of protecting the environment had been launched and supported in a number of countries, yet this never resulted in the practical, effective implementation of environmental policy by governments. In 2015, the Conference of Parties, COP 21, held in Paris brought together 195 countries from around the world. Subsequently, an agreement was signed by all participating countries, which stipulates that the increase in the Earth's global temperature will be limited to 2°C by the end of the 21st century. During this conference, surveys conducted made it possible to highlight the concerns of societies. In order of importance, the issues raised are categorized as follows: 1) energy; 2) water; 3) food; 4) environment; 5) poverty; 6) war; 7) disease; 8) education; 9) democracy and 10) populations.

However one might interpret this result, it is apparent that the issue of the environment is important to people and should be considered in all areas of research. Producing organic devices to provide and improve the comforts of day-to-day life is one of the objectives of the electronics industry, and in this approach, the Brabec triangle indicates the criteria for achieving, testing and marketing them.

Taking into account the evolution of our environment and keeping in mind its fragility and that it must be preserved for the future, the environmental dimension should be added to the criteria set out by Brabec. It would also occupy a key position in the hierarchy of criteria because, as we have analyzed, it would be of no use to manufacture new devices to replace existing ones where the energy expenditure and pollution generated exceed that necessary for the manufacture of existing devices.

The revised Brabec triangle is shown in Figure 4.5.

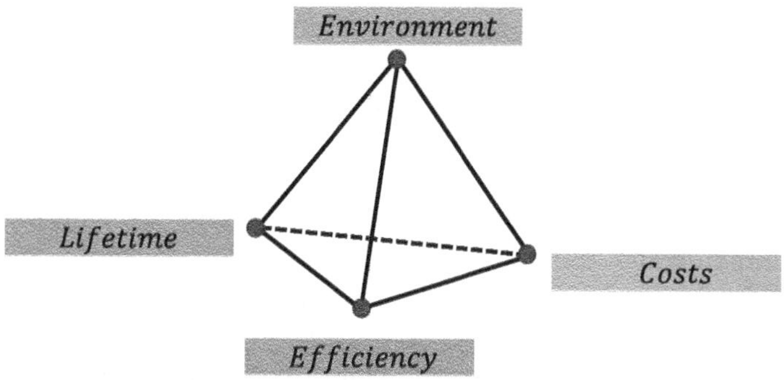

Figure 4.5. *Revised Brabec triangle. For a color version of this figure, see www.iste.co.uk/nguyen/electronics2.zip*

Currently, several commercial products resulting from organic electronics have already been available on the market in recent years. OLED screens are increasingly used to equip mobile telephones and other electronic devices such as televisions and computers. In lighting, OLED lamps for living rooms are available in different shapes and colors. They are also used to equip vehicles with optical accessories (on-board touch-sensor screen and car lights).

Organic solar cells are incorporated into handbags or backpacks and are used to charge small portable devices. They also equip bus shelters, making it possible to power small information screens.

Other uses of organic devices can be found, for example, in the manufacture of lithium batteries to power portable devices or in RFID tags for product identification.

It should be noted that the applications marketed only concern low-power devices, whereas they are non-existent for medium- and high-power devices due to the major obstacles of loss of efficiency and acceleration in degradation.

According to the Organic and Printed Electronics Association (OAS) road map, developments in organic electronics for the period 2013–2023 will be concentrated primarily on rollable and non-rollable OLED screens used in portable devices (telephones and tablets) and OLEDs for lighting.

Second, solar cells, transistors and organic batteries will be developed using the printing technique to be integrated into various systems used for power supply, memory or energy storage. The key parameters for each device type, as well as their related problems, have been clearly identified.

For OLEDs, these are: lifetime, luminance cost ($/watt or $/lumen), efficacy (lumen/watt) and diode size (m^2). From a technological point of view, the manufacturing problems relate to internal quantum efficiency (amount of photons emitted/ampere), optical efficiency (amount of photons emitted in the external medium), the stability of the color emitted and the uniformity of the luminance of the emitting surface. To improve the quality of OLEDs intended for lighting, it will be necessary to:

– reinforce the encapsulation to protect the diode from moisture and oxygen;

– increase the optical coupling efficiency. Existing solutions (use of microlenses, and substrates with a high refractive index) are not appropriate for marketing;

– improve the emission of blue light, which is the least stable in the long term;

– reduce the production cost to the level of competing products in order to gain access to marketing.

For OSCs, the key parameters are: conversion efficiency, power supplied per year, lifetime (outdoors), optical transmission (for semi-transparent modules), (mass) production technique and production cost (in $\$/W_p$). Problems relating to manufacturing technology include the choice of electrode materials (notably TCOs), encapsulation materials and primarily photo-active organic materials (stability). Like OLEDs, OSC quality and performance can be improved by improving the cell lifetime and reducing production costs. In the medium and long term, the cells will be incorporated into the glazing of cars and buildings serving as a complementary source of power.

The arguments developed in this chapter focus on three types of devices that are most promising for the future of organic electronics, namely LEDs, OSCs and OFETs. It should not be forgotten that many other applications using organic materials and printing manufacturing techniques are also studied and some have even been marketed. However, in terms of Brabec's criteria, these only target a specific section of consumers as a whole, and therefore are less prominent than the devices cited. It is also clear that improvements made to the devices studied will also benefit other organic applications. The same is true of the environmental constraints that we have discussed.

In under 30 years, research work in the field of electronics has been successful in bringing significant results to the understanding of physical processes and in developing the technology for manufacturing materials and devices, enabling the emergence of new electronic products very different from those designed by inorganic or silicon electronics. All of these products have common features constituting their advantages, such as very thin thickness, a flexible medium and low-cost mass production. Provided that lifetime and efficiency performance are improved, the electronic device market is expected to grow on a large scale in the near future. Organic electronics will not compete with traditional electronics but will complement them by providing an additional ecological touch, enabling them to adapt to the society of tomorrow, who we are convinced will live in harmony with and show respect for the environment.

List of Acronyms

General terms

AFM	Atomic force microscopy
AMOLED	Active-matrix organic light-emitting diode
BHJ	Bulk heterojunction
CB	Conduction band
CELIV	Charge extraction by linearly increasing voltage
CHEMFET	Chemical field-effect transistor
CIE	*Commission internationale de l'éclairage* (International Commission on Illumination)
COP	Conference of the Parties
DLOS	Deep-level optical spectroscopy
DLTS	Deep-level transient spectroscopy
DOS	Density of states
EBL	Electron blocking layer
EPBT	Energy payback time

EQE	External quantum efficiency
ETL	Electron transport layer
ETM	Electron transport material
FET	Field-effect transistor
GDM	Gaussian disorder model
GHG	Greenhouse gas
GWP	Global warming potential
HB	High barrier
HBL	Hole blocking layer
HI	Hysteresis index
HOMO	Highest occupied molecular orbital
HTL	Hole transport layer
HTM	Hole transport material
ICT	Integer charge transfer
IEA	International Energy Agency
IGFET	Insulated-gate field-effect transistor
IoT	Internet of Things
IPCE	Incident photon-to-current efficiency
LCD	Liquid crystal display
LCOE	Levelized cost of energy
LUMO	Lowest unoccupied molecular orbital
MOCVD	Metalorganic chemical vapor deposition

NIR	Near infrared
OE-A	Organic and Printed Electronics Association
OFET	Organic field-effect transistor
OLED	Organic light-emitting diode
OLET	Organic light-emitting transistor
OPV	Organic photovoltaic
OSC	Organic solar cell
PCE	Power conversion efficiency
PMOLED	Passive-matrix organic light-emitting diode
Q-DLTS	Charge-based deep-level transient spectroscopy
RFID	Radio-frequency identification
SAM	Self-assembled monolayer
SC	Semiconductor
SCLC	Space-charge limited current
SCR	Space charge region
SED	Stretched exponential decay
SERS	Surface-enhanced Raman scattering
SVA	Solvent vapor annealing
TADF	Thermally activated delayed fluorescence
TCO	Transparent conducting oxides
TFL	Trap-filled limit
TFT	Thin-film transistor

TLC	Trapped-limited conduction
TOF	Time-of-flight
TOLED	Top-emitting organic light-emitting diode
TSC	Thermally stimulated current
TTA	Triplet–triplet annihilation
UHB	Ultra-high barrier
UPS	UV photoelectron spectroscopy
VB	Valence band
VOLET	Vertical organic light-emitting transistor
VRH	Variable-range hopping
WOLED	White organic light-emitting diode
WVTR	Water vapor transmission rate
XPS	X-ray photoelectron spectroscopy

Materials

Alq_3	Tris(8-hydroxyquinoline)aluminum (III)
AZO	Aluminum-doped zinc oxide
BCB	Benzocyclobutene
BCzVBi	4,4'-bis(9-ethyl-3-carbazovinylene'-1,1'-biphenyl
BDD	Benzo[1,2-c:4,5-c']dithiophene-4,8-dione
BDPPV	Benzodifurandione-based PPV
BDT	Benzo[1,2-B:4,5-B']dithiophene

C-545T	10-(2-benzo-thiazolyl)-1,1,7,7-tetramethyl-2,3,6,7-tetrahydro-1H,5H,11H-[l]benzo-pyrano[6,7,8-ij]quinolizin-11
CFC-12	Dichlorofluoromethane
CN-PPVp	Poly(cyano-phenylene vinylene)
CPB	4,4'-N,N'-dicarbazolyl biphenyl
CYEP	Cyanoethylpullulane
CYTOP	Poly(perfluoroethylene-co-butenyl vinyl ether)
DCJTB	4-(dicyanomethylene)-2-t-butyl-6-(1,1,7,7-tetramethyljulolidyl-9-enyl)-4H-pyran
DCJTI	4-(dicyanomethylene)-2-i-propyl-6-(1,1,7,7-tetramethyljulolidyl-9-enyl)-4H-pyran
DH-4T	α,ω-dihexyl-quaterthiophene
DH-6T	Dihexyl-sexithiophene
DHTBT	4,7-di(4-hexylthien-2-yl)-benzothiadiazole
DM	N2,N2',N7,N7'-tetrakis (9,9-dimethyl-9H-fluorene-2-yl)- N2,N2'N7,N7'-tetrakis (4-methoxy phenyl)-9,9'-spirobi[fluorene]-2,2',7,7'-tetraamine
DMF	N,n-dimethylformamide
DMSO	Dimethylsulfoxide
DMZ	Dimethylzinc
DPVBi	4,4-bis(2,2-diphenylvinyl)biphenyl
F8T2	Poly[9,9-dioctylfluorene-co-bithiophene]
FIrpic	Bis[2-(4,6-difluorophenyl)pyridinato-C2,N](picolinato)iridium (III)

FTO	Fluorine-doped tin oxide
GO	Graphene oxide
HMDS	Hexamethyldisilazane
IPA	Isopropanol
Ir(HFP)	Tris[2,5-bis-2'(9',9'-dihexylfluorene)pyridine] iridium (III)
ITO	Indium tin oxide
m-MTDATA	4,4',4"-tris[3-methyl-phenyl(phenyl)amino] triphenylamine
MAPI	Methylammonium lead iodide perovskite
MDMO-PPV	Poly(2-methoxy-5-(3',7'-dimethyl-octyloxy)-1,4-phenylene vinylene)
MEH-PPV	Poly[2-methoxy-5-(2-ethyl-hexyloxy)-1,4-phenylene-vinyl]
MWCNT	Multi-walled carbon nanotubes
NDI	Naphthalene diimide
OTS	Octadecyltrichlorosilane
P3HT	Poly(hexylthiophene)
PANI	Polyaniline
PAT	Poly(alkylthiophene)
PBDB-T	Poly[(2,6-(4,8-bis(5-(2-ethylhexyl)thiophen-2-yl)-benzo[1,2-b:4,5-b']dithiophene))-alt-(5,5-(1',3'-di-2-thienyl-5',7'-bis(2-ethylhexyl)benzo[1',2'-c:4',5'-c']dithiophene-4,8-dione))]

PBDTTT-C	Poly[4,8-bis-alkyloxybenzo[1,2-b:4,5-b']dithiophene-2,6-diyl-alt-4-(alkyl-1-one)thieno[3,4-b]thiophene-2,6-diyl]
PBTTT	Poly(2,5-bis(3-alkylthiophen-2-yl)thieno(3,2-b)thiophene)
Pc	Phthalocyanine
PCBM	[6,6]-phenyl-C61-butyric acid methyl ester
PCDTBT	Poly[N-9'-heptadecanyl-2,7-carbazole-alt-5,5-(4,7-di-2-thienyl-2',1',3'-benzo-thiadiazole)]
PCPDTBT	Poly[2,6-(4,4-bis-(2-ethylhexyl)-4H-cyclopenta[2,1-b;3,4-b']dithiophene)-alt-4,7-(2,1,3-benzothiadiazole)]
PDI	Perylene diimide
PDMS	Poly(dimethysiloxane)
PEDOT-PFESA	Poly(3,4-ethylene-dioxythiophene)-poly(perfluoroethylene sulfonic acid)
PEDOT:PSS	Poly(3,4-ethylenedioxythiophene)-poly(styrenesulfonate)
PET	Poly(ethylene terephthalate)
PFH-MEH	Poly[(9,9-dihexyloxyfluoren-2,7-diyl)-alt-co-(2-methoxy-5-{2-ethylhexyloxy}phenylen-1,4-diyl)]
PFO	Polyfluorene
PGMEA	Propylene glycol-monomethyl ether acetate
PI	Polyimide
PIF	Poly(3,9-di-*t*-butylindeno[1,2-b]fluorene)
PITN	Poly(isathianapthene)

PMMA	Poly(methyl methacrylate)
POSS	Polyhedral oligomeric silsesquioxane
PPP	Poly(p-phenylene)
PPV	Poly(phenylene vinylene)
PPy	Polypyrrole
PSBF	Poly(spirobifluorene)
PTAA	Poly(triaryl amine)
PTB7-TH	Poly[4,8-bis(5-(2-ethylhexyl)thiophen-2-yl)benzo[1,2-b;4,5-b']dithiophene-2,6-diyl-alt-(4-(2-ethylhexyl)-3-fluorothieno[3,4-b]thiophene-)-(2-carboxylate-2-6-diyl)]
PTQ	Poly[[6,7-difluoro[(2-hexyldecyl)oxy]-5,8-quinoxalinediyl]-2,5-thiophenediyl]
PVA	Poly(vinyl alcohol)
PVK	Poly(n-vinylcarbazole)
PVP	Poly-4-vinylphenol
SO	Dibenzothiophene-*S,S*-dioxide
Spiro-OMeTAD	2,2',7,7'-tetrakis-(N,N-di-4-methoxyphenylamino)-9,9'-spirobifluorene
TBADN	2-tert-butyl-9,10-di(naphth-2-yl)anthracene
Tips-pentacene	6,13-bis (triisopropyl-silylethynyl)pentacene
TPA	Triphenylamine

References

Ahn, J.H., Wang, C., Widdowson, N.E., Pearson, C., Bryce, M.R., Petty, M.C. (2005). Thermal annealing of blended-layer organic light-emitting diodes. *Journal of Applied Physics*, 98, 054508.

Albrecht, S., Saliba, M., Correa Baena, J.P., Lang, F., Kegelmann, L., Mews, M., Steier, L., Abate, A., Rappich, J., Korte, L., Schlatmann, R., Nazeeruddin, M.K., Hagfeldt, A., Grätzel, M., Rech, B. (2016). Monolithic perovskite/silicon-heterojunction tandem solar cells processed at low temperature. *Energy and Environmental Science*, 9(1), 81–88.

Aziz, H., Popovic, Z.D., Hu, N.X. (2002). Organic light emitting devices with enhanced operational stability at elevated temperatures. *Applied Physics Letters*, 81(2), 370–372.

Bailie, C.D., Christoforo, M.G., Mailoa, J.P., Bowring, A.R., Unger, E.L., Nguyen, W.H., Burschka, J., Pellet, N., Lee, J.Z., Grätzel, M., Noufi, R., Buonassisi, T., Salleo, A., McGehee, M.D. (2015). Semi-transparent perovskite solar cells for tandems with silicon and CIGS. *Energy and Environmental Science*, 8(3), 956–963.

Bao, Z. and Locklin, J. (2007). *Organic Field-Effect Transistors*. CRC Press, Boca Raton.

Bisquert, J. (2018). *The Physics of Solar Cells*. CRC Press, Boca Raton.

Bobbert, P.A., Sharma, A., Mathijssen, S.G.J., Kemerink, M., de Leeuw, D.M. (2012). Operational stability of organic field-effect transistors. *Advanced Materials*, 24(9), 1146–1158.

Brabec, C.J. (2004). Organic photovoltaics: Technology and market. *Solar Energy Materials and Solar Cells*, 83(2/3), 273–292.

Brabec, C.J., Hauch, J.A., Schilinsky, P., Waldauf, C. (2005). Production aspects of organic photovoltaics and their impact on the commercialization of devices. *MRS Bulletin*, 30(1), 50–52.

Burke, D.J. and Lipomi, D.J. (2013). Green chemistry for organic solar cells. *Energy and Environmental Science*, 6(7), 2053–2066.

Capelli, R., Toffanin, S., Generali, G., Usta, H., Facchetti, A., Muccini, M. (2010). Organic light-emitting transistors with an efficiency that outperforms the equivalent light-emitting diodes. *Nature Materials*, 9(6), 496–503.

Chang, H.W., Lee, J., Hofmann, S., Hyun Kim, Y., Müller-Meskamp, L., Lüssem, B., Wu, C.C., Leo, K., Gather, M.C. (2013). Nano-particle based scattering layers for optical efficiency enhancement of organic light-emitting diodes and organic solar cells. *Journal of Applied Physics*, 113(20).

Chen, F.C., Chu, C.W., He, J., Yang, Y., Lin, J.L. (2004). Organic thin-film transistors with nanocomposite dielectric gate insulator. *Applied Physics Letters*, 85(15), 3295–3297.

Chen, H.W., Sakai, N., Ikegami, M., Miyasaka, T. (2015). Emergence of hysteresis and transient ferroelectric response in organo-lead halide perovskite solar cells. *Journal of Physical Chemistry Letters*, 6(1), 164–169.

Cheng, Y.J., Yang, S.H., Hsu, C.S. (2009). Synthesis of conjugated polymers for organic solar cell applications. *Chemical Reviews*, 109(11), 5868–5923

Chirvase, D., Parisi, J., Hummelen, J.C., Dyakonov, V. (2004). Influence of nanomorphology on the photovoltaic action of polymer–fullerene composites. *Nanotechnology*, 15(9), 1317–1323.

Chiu, S.W., Lin, L.Y., Lin, H.W., Chen, Y.H., Huang, Z.Y., Lin, Y.T., Lin, F., Liu, Y.H., Wong, K.T. (2012). A donor–acceptor–acceptor molecule for vacuum-processed organic solar cells with a power conversion efficiency of 6.4%. *Chemical Communications*, 48(13), 1857–1859.

Cho, Y.J., Yook, K.S., Lee, J.Y. (2014). A universal host material for high external quantum efficiency close to 25% and long lifetime in green fluorescent and phosphorescent OLEDs. *Advanced Materials*, 26(24), 4050–4055.

Chua, L.L., Zaumseil, J., Chang, J.F., Ou, E.C.W.N., Ho, P.K.M., Sirringhaus, H., Friend, R.H. (2005). General observation of n-type field-effect behaviour in organic semiconductors. *Nature*, 434, 194–199.

Collet, J., Tharaud, O., Chapoton, A., Vuillaume, D. (2000). Low-voltage, 30 nm channel length, organic transistors with a self-assembled monolayer as gate insulating films. *Applied Physics Letters*, 76(14), 1941–1943.

Crone, B., Dodabalapur, A., Gelperin, A., Torsi, L., Katz, H.E., Lovinger, A.J., Bao, B. (2001). Electronic sensing of vapors with organic transistors. *Applied Physics Letters*, 78(15), 2229–2231.

Cui, Y., Yao, H., Hong, L., Zhang, T., Xu, Y., Xian, K., Gao, B., Qin, J., Zhang, J., Wei, Z., Hou, J. (2019). Achieving over 15% efficiency in organic photovoltaic cells via copolymer design. *Advanced Materials*, 31(14), 1–7.

Cumpston, B.H., Parker, I.D., Jensen, K.F. (1997). *In situ* characterization of the oxidative degradation of a polymeric light emitting device. *Journal of Applied Physics*, 81(8), 3716–3720.

Darling, S.B. and You, F. (2013). The case for organic photovoltaics. *RSC Advances*, 3(39), 17633–17648.

Deibel, C. and Dyakonov, V. (2010). Polymer fullerene bulk heterojunction solar cells. *Report on Progress in Physics*, 73.

Dennler, G., Prall, H.J., Koeppe, R., Egginger, M., Autengruber, R., Sariciftci, N.S. (2006). Enhanced spectral coverage in tandem organic solar cells. *Applied Physics Letters*, 89(7), 1–4.

Di, C.A., Yu, G., Liu, Y., Guo, Y., Sun, X., Zheng, J., Wen, Y., Wu, W., Zhu, D. (2009). Selective crystallization of organic semiconductors for high performance organic field-effect transistors. *Chemistry of Materials*, 21(20), 4873–4879.

Dinelli, F., Capelli, R., Loi, M.A., Murgia, M., Muccini, M., Facchetti, A., Marks, T.J. (2006). High-mobility ambipolar transport in organic light-emitting transistors. *Advanced Materials*, 18(11), 1416–1420.

Duggan, A.R. (2005). Organic electroluminescent devices for solid state lighting. In *Organic Electroluminescence*, Kafafi, Z. (ed.). CRC Press, Boca Raton.

Egginger, M., Bauer, S., Schwödiauer, R., Neugebauer, H., Sariciftci, N.S. (2009). Current versus gate voltage hysteresis in organic field effect transistors. *Monatshefte fur Chemie*, 140(7), 735–750.

Erb, T., Zhokhavets, U., Gobsch, G., Raleva, S., Stühn, B., Schilinsky, P., Waldauf, C., Brabec, C.J. (2005). Correlation between structural and optical properties of composite polymer/fullerene films for organic solar cells. *Advanced Functional Materials*, 15(7), 1193–1196.

Facchetti, A. (2007). Semiconductors for organic transistors. *Materials Today*, 10(3), 28–37.

Facchetti, B.A., Mushrush, M., Katz, H.E., Marks, T.J. (2003). n-Type building blocks for organic electronics. *Advanced Materials*, 15(1), 33–38.

Féry, C., Racine, B., Vaufrey, D., Doyeux, H., Cinà, S. (2005). Physical mechanism responsible for the stretched exponential decay behavior of aging organic light-emitting diodes. *Applied Physics Letters*, 87(21), 1–3.

Filipic, M., Loper, P., Niesen, B., De Wolf, S., Krc, J., Ballif, C., Topic, M. (2015). CH3NH3PbI3 perovskite/silicon tandem solar cells: Characterization based optical simulations. *Optic Express*, 23, A263–A278.

Futscher, M.H. and Ehrler, B. (2016). Efficiency limit of perovskite/Si tandem solar cells. *ACS Energy Letters*, 1(4), 863–868.

Gambhir, A., Sandwell, P., Nelson, J. (2016). The future costs of OPV – A bottom-up model of material and manufacturing costs with uncertainty analysis. *Solar Energy Materials and Solar Cells*, 156, 49–58.

Garnier, F., Yassar, A., Hajlaoui, R., Horowitz, G., Deloffre, F., Servet, B., Ries, S., Alnot, P. (1993). Molecular engineering of organic semiconductors: Design of self-assembly properties in conjugated thiophene oligomers. *Journal of the American Chemical Society*, 115(19), 8716–8721.

Gevorgyan, S.A., Madsen, M.V., Dam, H.F., Jørgensen, M., Fell, C.J., Anderson, K.F., Duck, B.C., Mescheloff, A., Katz, E.A., Elschner, A., Roesch, R., Hoppe, H., Hermenau, M., Riede, M., Krebs, F.C. (2013). Interlaboratory outdoor stability studies of flexible roll-to-roll coated organic photovoltaic modules: Stability over 10,000 h. *Solar Energy Materials and Solar Cells*, 116, 187–196.

Giorgi, G., Fujisawa, J.I., Segawa, H., Yamashita, K. (2013). Small photocarrier effective masses featuring ambipolar transport in methylammonium lead iodide perovskite: A density functional analysis. *Journal of Physical Chemistry Letters*, 4(24), 4213–4216.

Gong, X., Moses, D., Heeger, A.J. (2004). White electrophosphorescence from semi-conducting polymer blends. *Organic Light-Emitting Materials and Devices VIII*, 5519(7), 82.

Gu, G., Bulović, V., Burrows, P.E., Forrest, S.R., Thompson, M.E. (1996). Transparent organic light emitting devices. *Applied Physics Letters*, 68(19), 2606–2608.

Gundlach, D.J., Lin, Y.Y., Jackson, T.N., Schlom, D.G. (1997). Oligophenyl-based organic thin film transistors. *Applied Physics Letters*, 71(26), 3853–3855.

Gunes, S., Neugebauer, H., Sariciftci, N.S. (2007). Conjugated polymer-based organic solar cells, *Chem. Rev.*, 107(4), 1324–1338.

Guo, Y., Yu, G., Liu, Y. (2010). Functional organic field-effect transistors. *Advanced Materials*, 22(40), 4427–4447.

Hailegnaw, B., Kirmayer, S., Edri, E., Hodes, G., Cahen, D. (2015). Rain on methylammonium lead iodide based perovskites: Possible environmental effects of perovskite solar cells. *Journal of Physical Chemistry Letters*, 6(9), 1543–1547.

Hammock, M.L., Chortos, A., Tee, B.C.K., Tok, J.B.H., Bao, Z. (2013). 25th anniversary article: The evolution of electronic skin (E-Skin): A brief history, design considerations, and recent progress. *Advanced Materials*, 25(42), 5997–6038.

Han, S.H., Kim, J.H., Jang, J., Cho, S.M., Oh, M.H., Lee, S.H., Choo, D.J. (2006). Lifetime of organic thin-film transistors with organic passivation layers. *Applied Physics Letters*, 88(7), 86–89.

Handy, R.D., Owen, R., Valsami-Jones, E. (2008). The ecotoxicology of nanoparticles and nanomaterials: Current status, knowledge gaps, challenges, and future needs. *Ecotoxicology*, 17(5), 315–325.

Herwig, P.T. and Müllen, K. (1999). Soluble pentacene precursor: Synthesis, solid-state conversion into pentacene and application in a field-effect transistor. *Advanced Materials*, 11(6), 480–483.

Hoppe, H., Niggemann, M., Winder, C., Kraut, J., Hiesgen, R., Hinsch, A., Meissner, D., Sariciftci, N.S. (2004). Nanoscale morphology of conjugated polymer/fullerene-based bulk-heterojunction solar cells. *Advanced Functional Materials*, 14(10), 1005–1011.

Horowitz, B.G. (1999). Organic field-effect transistors. *Advanced Materials*, 10(5), 365–377.

Horowitz, B.G. (2004). Organic thin film transistors: From theory to real devices. *Journal of Materials Research*, 19(7), 1946–1962.

Hsieh, H.C., Hsiow, C.Y., Lin, K.F., Shih, Y.C., Wang, L., Renaud, C., Nguyen, T.P. (2018). Analysis of defects and traps in N-I-P layered-structure of perovskite solar cells by charge-based deep level transient spectroscopy (Q-DLTS). *Journal of Physical Chemistry C*, 122(31), 17601–17611.

Huang, J., Li, G., Wu, E., Xu, Q., Yang, Y. (2006). Achieving high-efficiency polymer white-light-emitting devices. *Advanced Materials*, 18(1), 114–117.

Hwang, D.K., Fuentes-Hernandez, C., Kim, J., Potscavage, W.J., Kim, S.J., Kippelen, B. (2011). Top-gate organic field-effect transistors with high environmental and operational stability. *Advanced Materials*, 23(10), 1293–1298.

Ishii, M. and Taga, Y. (2002). Influence of temperature and drive current on degradation mechanisms in organic light-emitting diodes. *Applied Physics Letters*, 80(18), 3430–3432.

Janata, J. and Josowicz, M. (2003). Conducting polymers in electronic chemical sensors. *Nature Materials*, 2(1), 19–24.

Janssen, R.A.J. and Nelson, J. (2013). Factors limiting device efficiency in organic photovoltaics. *Advanced Materials*, 25(13), 1847–1858.

Jeon, N.J., Noh, J.H., Kim, Y.C., Yang, W.S., Ryu, S., Seok, S.I. (2014). Solvent engineering for high-performance inorganic-organic hybrid perovskite solar cells. *Nature Materials*, 13(9), 897–903.

Jeon, N.J., Na, H., Jung, E.H., Yang, T.Y., Lee, Y.G., Kim, G., Shin, H.W., Seok, S.I., Lee, J., Seo, J. (2018). A fluorene-terminated hole-transporting material for highly efficient and stable perovskite solar cells. *Nature Energy*, 3(8), 682–689.

Jiang, Q., Chu, Z., Wang, P., Yang, X., Liu, H., Wang, Y., Yin, Z., Wu, J., Zhang, X., You, J. (2017). Planar-structure perovskite solar cells with efficiency beyond 21%. *Advanced Materials*, 29(46), 1–7.

Kaji, H., Suzuki, H., Fukushima, T., Shizu, K., Suzuki, K., Kubo, S., Komino, T., Oiwa, H., Suzuki, F., Wakamiya, A., Murata, Y., Adachi, C. (2015). Purely organic electro-luminescent material realizing 100% conversion from electricity to light. *Nature Communications*, 6(May), 2–9.

Kalinowski, J. (2005). Emission mechanisms in organic light-emitting diodes. *Organic Electroluminescence*, 2005, 23–112.

Kalinowski, J., Cocchi, M., Virgili, D., Fattori, V., Williams, J.A.G. (2007). Mixing of excimer and exciplex emission: A new way to improve white light emitting organic electrophosphorescent diodes. *Advanced Materials*, 19(22), 4000–4005.

Kalowekamo, J. and Baker, E. (2009). Estimating the manufacturing cost of purely organic solar cells. *Solar Energy*, 83(8), 1224–1231.

Kato, K., Iwasaki, T., Tsujimura, T. (2015). Over 130 lm/W all-phosphorescent white OLEDs for next-generation lighting. *Journal of Photopolymer Science and Technology*, 28(3), 335–340.

Kehrer, L.A., Winter, S., Fischer, R., Melzer, C., Von Seggern, H. (2012). Temporal and thermal properties of optically induced instabilities in P3HT field-effect transistors. *Synthetic Metals*, 161(23/24), 2558–2561.

Keskinen, M. and Valkama, J. (2009). End-of-life challenges of printed electronics. *2009 IEEE International Symposium on Sustainable Systems and Technology*, 18–19 May.

Kim, H.S. and Park, N.G. (2014). Parameters affecting I-V hysteresis of $CH_3NH_3PbI_3$ perovskite solar cells: Effects of perovskite crystal size and mesoporous TiO_2 layer. *Journal of Physical Chemistry Letters*, 5(17), 2927–2934.

Kim, J.S., Ho, P.K.H., Greenham, N.C., Friend, R.H. (2000). Electroluminescence emission pattern of organic light-emitting diodes: Implications for device efficiency calculations. *Journal of Applied Physics*, 88(2), 1073–1081.

Kirchartz, T., Taretto, K., Rau, U. (2009). Efficiency limits of organic bulk heterojunction solar cells. *Journal of Physical Chemistry C*, 113(41), 17958–17966.

Koh, T.W., Choi, J.M., Lee, S., Yoo, S. (2010). Optical outcoupling enhancement in organic light-emitting diodes: Highly conductive polymer as a low-index layer on microstructured ITO electrodes. *Advanced Materials*, 22(16), 1849–1853.

Kojima, A., Teshima, K., Shirai, Y., Miyasaka, T. (2009). Organometal halide perovskites as visible-light sensitizers for photovoltaic cells. *Journal of the American Chemical Society*, 131(17), 6050–6051.

Komoda, T., Yamae, K., Kittichungchit, V. (2012). Extremely high performance white OLEDs for lighting. *SID's Digest of Technical Papers*, 43(1), 610–613.

Kondakov, D.Y. (2015). Triplet–triplet annihilation in highly efficient fluorescent organic light-emitting diodes: Current state and future outlook. *Philosophical Transactions of the Royal Society A: Mathematical, Physical and Engineering Sciences*, 373, 2140321.

Krebs, F.C. (2008). *Polymer Photovoltaic*. SPIE Press, Bellingham.

Krebs, F.C. and Norrman, K. (2007). Analysis of the failure mechanism for a stable organic photovoltaic during 10 000 h of testing. *Progress in Photovoltaics: Research and Applications*, 15, 697–712.

Krebs, F.C. and Spanggaard, H. (2005). Significant improvement of polymer solar cell stability. *Chemistry of Materials*, 17(21), 5235–5237.

Krishnamoorthy, T., Ding, H., Yan, C., Leong, W.L., Baikie, T., Zhang, Z., Sherburne, M., Li, S., Asta, M., Mathews, N., Mhaisalkar, S.G. (2015). Lead-free germanium iodide perovskite materials for photovoltaic applications. *Journal of Materials Chemistry A*, 3(47), 23829–23832.

Kunnari, E., Valkama, J., Keskinen, M., Mansikkamäki, P. (2009). Environmental evaluation of new technology: Printed electronics case study. *Journal of Cleaner Production*, 17(9), 791–799.

Lee, J.H., Wu, C.I., Liu, S.W., Huang, C.A., Chang, Y. (2005). Mixed host organic light-emitting devices with low driving voltage and long lifetime. *Applied Physics Letters*, 86(10), 1–3.

Lee, M.M., Teuscher, J., Miyasaka, T., Murakami, T.N., Snaith, H.J. (2012). Efficient hybrid solar cells based on meso-superstructured organometal halide perovskites. *Science*, 338(6107), 643–647.

Lee, C., Kang, D.J., Kang, H., Kim, T., Park, J., Lee, J., Yoo, S., Kim, B.J. (2014). Simultaneously enhancing light extraction and device stability of organic light-emitting diodes using a corrugated polymer nanosphere templated PEDOT:PSS layer. *Advanced Energy Materials*, 4(8), 1–6.

Lee, H.K.H., Li, Z., Durrant, J.R., Tsoi, W.C. (2016). Is organic photovoltaics promising for indoor applications? *Applied Physics Letters*, 108, 253301.

Lei, T., Dou, J.H., Cao, X.Y., Wang, J.Y., Pei, J. (2013). Electron-Deficient Poly(p-phenylene vinylene) provides electron mobility over 1 cm^2 V^{-1} s^{-1} under ambient conditions. *Journal of the American Chemical Society*, 135(33), 12168–12171.

Leijtens, T., Eperon, G.E., Pathak, S., Abate, A., Lee, M.M., Snaith, H.J. (2013). Overcoming ultraviolet light instability of sensitized TiO_2 with meso-superstructured organometal tri-halide perovskite solar cells. *Nature Communications*, 4, 1–8.

Leijtens, T., Stranks, S.D., Eperon, G.E., Lindblad, R., Johansson, E.M.J., McPherson, I.J., Rensmo, H., Ball, J.M., Lee, M.M., Snaith, H.J. (2014). Electronic properties of meso-superstructured and planar organometal halide perovskite films: Charge trapping, photodoping, and carrier mobility. *ACS Nano*, 8(7), 7147–7155.

Li, G., Shrotriya, V., Huang, J., Moriarty, T., Emery, K., Yang, Y. (2005a). High-efficiency solution processable polymer photovoltaic cells by self-organization of polymer blends. *Nature Materials*, 4, 864–868.

Li, Y., Rizzo, A., Mazzeo, M., Carbone, L., Manna, L., Cingolani, R., Gigli, G. (2005b). White organic light-emitting devices with CdSe/ZnS quantum dots as a red emitter. *Journal of Applied Physics*, 97 113501.

Li, Z., He, G., Wan, X., Liu, Y., Zhou, J., Long, G., Zuo, Y., Zhang, M., Chen, Y. (2012). Solution processable rhodanine-based small molecule organic photovoltaic cells with a power conversion efficiency of 6.1%. *Advanced Energy Materials*, 2(1), 74–77.

Liew, Y.F., Aziz, H., Hu, N.X., Chan, H.S.O., Xu, G., Popovic, Z. (2000). Investigation of the sites of dark spots in organic light-emitting devices. *Applied Physics Letters*, 77(17), 2650–2652.

Lim, S.F., Wang, W., Chua, S.J. (2001). Degradation of organic light-emitting devices due to formation and growth of dark spots. *Materials Science and Engineering B: Solid-State Materials for Advanced Technology*, 85(2/3), 154–159.

Lin, Y., Gundlach, D., Nelson, S., Jackson, T. (1997). Stacked pentacene layer organic thin-film transistors with improved characteristics. *IEEE Electron Device Letters*, 18(12), 606–608.

Liu, M., Johnston, M.B., Snaith, H.J. (2013). Efficient planar heterojunction perovskite solar cells by vapour deposition. *Nature*, 501(7467), 395–398.

Liu, C., Yi, C., Wang, K., Yang, Y., Bhatta, R.S., Tsige, M., Xiao, S., Gong, X. (2015). Single-junction polymer solar cells with over 10% efficiency by a novel two-dimensional donor-acceptor conjugated copolymer. *ACS Applied Materials and Interfaces*, 7(8), 4928–4935.

Lu, L., Zheng, T., Wu, Q., Schneider, A.M., Zhao, D., Yu, L. (2015). Recent advances in bulk heterojunction polymer solar cells. *Chemical Reviews*, 115(23), 12666–12731.

Ma, L. and Yang, Y. (2004). Unique architecture and concept for high-performance organic transistors. *Applied Physics Letters*, 85(21), 5084–5086.

Ma, W., Yang, C., Gong, X., Lee, K., Heeger, A.J. (2005). Thermally stable, efficient polymer solar cells with nanoscale control of the interpenetrating network morphology. *Advanced Functional Materials*, 15(10), 1617–1622.

Madigan, C.F., Lu, M.H., Sturm, J.C. (2000). Improvement of output coupling efficiency of organic light-emitting diodes by backside substrate modification. *Applied Physics Letters*, 76(13), 1650–1652.

Manceau, M., Rivaton, A., Gardette, J.L., Guillerez, S., Lemaître, N. (2009). The mechanism of photo- and thermooxidation of poly(3-hexylthiophene) (P3HT) reconsidered. *Polymer Degradation and Stability*, 94(6), 898–907.

Mathijssen, S.G.J., Cölle, M., Gomes, H., Smits, E.C.P., De Boer, B., McCulloch, I., Bobbert, P.A., de Leeuw, D.M. (2007). Dynamics of threshold voltage shifts in organic and amorphous silicon field-effect transistors. *Advanced Materials*, 19(19), 2785–2789.

McCulloch, I., Heeney, M., Bailey, C., Genevicius, K., MacDonald, I., Shkunov, M., Sparrowe, D., Tierney, S., Wagner, R., Zhang, W., Chabinyc, M.L., Kline, R.J., McGehee, M.D., Toney, M.F. (2006). Liquid-crystalline semiconducting polymers with high charge-carrier mobility. *Nature Materials*, 5(4), 328–333.

Meijer, E.J., de Leeuw, D.M., Setayesh, S., Van Veenendaal, E., Huisman, B.H., Blom, P.W.M., Hummelen, J.C., Scherf, U., Klapwijk, T.M. (2003). Solution-processed ambipolar organic field-effect transistors and inverters. *Nature Materials*, 2(10), 678–682.

Meng, L., Zhang, Y., Wan, X., Li, C., Zhang, X., Wang, Y., Ke, X., Xiao, Z., Ding, L., Xia, R., Yip, H.L., Cao, Y., Chen, Y. (2018). Organic and solution-processed tandem solar cells with 17.3% efficiency. *Science*, 361(6407), 1094–1098.

Nakamura, S. (1998). The roles of structural imperfections in InGaN-based blue light-emitting diodes and laser diodes. *Science*, 281, 956–961.

Nakamura, T., Tsutsumi, N., Juni, N., Fujii, H. (2005). Thin-film waveguiding mode light extraction in organic electroluminescent device using high refractive index substrate. *Journal of Applied Physics*, 97, 054505.

National Renewable Energy Laboratory (2017). Research cell record efficiency chart. Report, NREL.

Nayak, P.K., Mahesh, S., Snaith, H.J., Cahen, D. (2019). Photovoltaic solar cell technologies: Analysing the state of the art. *Nature Reviews Materials*, 4(4), 269–285.

Newsome, C.J., Kawase, T., Shimoda, T., Brennan, D.J. (2003). Phase behavior of polymer semiconductor films and its influence on the mobility in FET devices. *Organic Field Effect Transistors II*, 5217, 16.

Nguyen, T.P. and Renaud, C. (2009). Interface effects on the defect state formation in organic devices. *Organic Photonic Materials and Devices XI*, 7213.

Nguyen, T.P. and Yang, S.H. (2018). Hybrid materials based on polymer nano-composites for environmental applications. In *Polymer-based Nano-composites for Energy and Environmental Applications*, Jawaid, M. and Khan, M.M. (eds). Woodhead Publishing, Sawston.

Nguyen, T.P., Ip, J., Jolinat, P., Destruel, P. (2001). XPS and sputtering study of the Alq3/electrode interfaces in organic light emitting diodes. *Applied Surface Science*, 172(1/2), 75–83.

Nguyen, T.P., Ip, J., Renaud, C., Huang, C.H., Guillen, C., Herrero, J. (2006). Study of the interface formed between poly(2-methoxy-5-(2′-ethyl-hexyloxyl)-p-phenylene vinylene) and indium tin oxide in top emission organic light emitting diodes. *Applied Surface Science*, 252(24), 8388–8393.

Nguyen, T.P., Girault, P., Renaud, C., Reisdorffer, F., Le Rendu, P., Wang, L. (2014). Effects of the negative electrode contact on the performance of poly(hexylthiophene):6,6-phenyl-C61-butyric acid methyl ester based organic solar cells. *Journal of Applied Physics*, 115(1), 1–8.

Niu, G., Guo, X., Wang, L. (2015). Review of recent progress in chemical stability of perovskite solar cells. *Journal of Materials Chemistry A*, 3(17), 8970–8980.

Noel, N.K., Stranks, S.D., Abate, A., Wehrenfennig, C., Guarnera, S., Haghighirad, A.A., Sadhanala, A., Eperon, G.E., Pathak, S.K., Johnston, M.B., Petrozza, A., Herz, L.M., Snaith, H.J. (2014). Lead-free organic-inorganic tin halide perovskites for photovoltaic applications. *Energy and Environmental Science*, 7(9), 3061–3068.

Ong, B.S., Wu, Y., Liu, P., Gardner, S. (2004). High-performance semiconducting polythiophenes for organic thin-film transistors. *Journal of the American Chemical Society*, 126(11), 3378–3379.

Park, S.Y., Park, M., Lee, H.H. (2004). Cooperative polymer gate dielectrics in organic thin-film transistors. *Applied Physics Letters*, 85(12), 2283–2285.

Park, S.K., Jackson, T.N., Anthony, J.E., Mourey, D.A. (2007). High mobility solution processed 6,13-bis(triisopropyl-silylethynyl) pentacene organic thin film transistors. *Applied Physics Letters*, 91(6), 6–9.

Pellet, N., Gao, P., Gregori, G., Yang, T.Y., Nazeeruddin, M.K., Maier, J., Grätzel, M. (2014). Mixed-organic-cation perovskite photovoltaics for enhanced solar-light harvesting. *Angewandte Chemie – International Edition*, 53(12), 3151–3157.

Peumans, P., Yakimov, A., Forrest, S.R. (2003). Small molecular weight organic thin-film photodetectors and solar cells. *Journal of Applied Physics*, 93(7), 3693–3723.

Ramírez Quiroz, C.O., Shen, Y., Salvador, M., Forberich, K., Schrenker, N., Spyropoulos, G.D., Heumüller, T., Wilkinson, B., Kirchartz, T., Spiecker, E., Verlinden, P.J., Zhang, X., Green, M.A., Ho-Baillie, A., Brabec, C.J. (2018). Balancing electrical and optical losses for efficient 4-terminal Si-perovskite solar cells with solution processed percolation electrodes. *Journal of Materials Chemistry A*, 6(8), 3583–3592.

Reese, M.O., Gevorgyan, S.A., Jørgensen, M., Bundgaard, E., Kurtz, S.R., Ginley, D.S., Olson, D.C., Lloyd, M.T., Morvillo, P., Katz, E.A. *et al.* (2011). Consensus stability testing protocols for organic photovoltaic materials and devices. *Solar Energy Materials and Solar Cells*, 95(5), 1253–1267.

Reineke, S., Thomschke, M., Lüssem, B., Leo, K. (2013). White organic light-emitting diodes: Status and perspective. *Reviews of Modern Physics*, 85(3), 1245–1293.

Reisdorffer, F., Haas, O., Le Rendu, P., Nguyen, T.P. (2012). Co-solvent effects on the morphology of P3HT:PCBM thin films. *Synthetic Metals*, 161(23/24), 2544–2548.

Richards, T. and Sirringhaus, H. (2008). Bias-stress induced contact and channel degradation in staggered and coplanar organic field-effect transistors. *Applied Physics Letters*, 92(2), 90–93.

Ricks, M.L., Vargas, J.R., Klubek, K.P., Jarikov, V.V., Liao, L.-S., Helber, M.J., Begley, W.J., Hatwar, T.K., Conley, S.R., Cosimbescu, L., Brown, C.T. (2007). P-169: Efficient, long-lifetime OLED host and dopant formulations for full-color displays. *SID's Digest of Technical Papers*, 38(1), 830–833.

Riel, H., Karg, S., Beierlein, T., Rieß, W., Neyts, K. (2003). Tuning the emission characteristics of top-emitting organic light-emitting devices by means of a dielectric capping layer: An experimental and theoretical study. *Journal of Applied Physics*, 94(8), 5290–5296.

Roes, A.L., Alsema, E.A., Blok, K., Patel, M.K. (2009). Ex-ante environmental and economic evaluation of polymer. *Progress in Photovoltaics: Research and Applications*, 17, 372–393.

Roncali, J. (1997). Synthetic principles for bandgap control in linear π-conjugated systems. *Chemical Reviews*, 97(1), 173–205.

Ryu, K., Kymissis, I., Bulović, V., Sodini, C.G. (2005). Direct extraction of mobility in pentacene OFETs using C-V and I-V measurements. *IEEE Electron Device Letters*, 26(10), 716–718.

Sacco, A., Rolle, L., Scaltrito, L., Tresso, E., Pirri, C.F. (2013). Characterization of photovoltaic modules for low-power indoor application. *Applied Energy*, 102, 1295–1302.

Sakai, J., Taima, T., Yamanari, T., Saito, K. (2009). Annealing effect in the sexithiophene:C70 small molecule bulk heterojunction organic photovoltaic cells. *Solar Energy Materials and Solar Cells*, 93(6/7), 1149–1153.

Saliba, M., Matsui, T., Seo, J.Y., Domanski, K., Correa-Baena, J.P., Nazeeruddin, M.K., Zakeeruddin, S.M., Tress, W., Abate, A., Hagfeldt, A., Grätzel, M. (2016). Cesium-containing triple cation perovskite solar cells: Improved stability, reproducibility and high efficiency. *Energy and Environmental Science*, 9(6), 1989–1997.

Santato, C., Cicoira, F., Martel, R. (2011). Organic photonics: Spotlight on organic transistors. *Nature Photonics*, 5(7), 392–393.

Sariciftci, N.S., Smilowitz, L., Heeger, A.J., Wudi, F. (1992). Photoinduced electron transfer from a conducting polymer to buckminsterfullerene. *Science*, 1992, 3885–3887.

Sasabe, H. and Kido, J. (2013). Development of high performance OLEDs for general lighting. *Journal of Materials Chemistry C*, 1699–1707.

Scharber, M.C., Mühlbacher, D., Koppe, M., Denk, P., Waldauf, C., Heeger, A.J., Brabec, C.J. (2006). Design rules for donors in bulk-heterojunction solar cells – Towards 10% energy-conversion efficiency. *Advanced Materials*, 18(6), 789–794.

Schilinsky, P., Waldauf, C., Brabec, C.J. (2002). Recombination and loss analysis in polythiophene-based bulk heterojunction photodetectors. *Applied Physics Letters*, 81(20), 3885–3887.

Schmidt, G.C., Bellmann, M., Meier, B., Hambsch, M., Reuter, K., Kempa, H., Hübler, A.C. (2010). Modified mass printing technique for the realization of source/drain electrodes with high resolution. *Organic Electronics: Physics, Materials, Applications*, 11(10), 1683–1687.

Scholz, S., Kondakov, D., Lüssem, B., Leo, K. (2015). Degradation mechanisms and reactions in organic light-emitting devices. *Chemical Reviews*, 115(16), 8449–8503.

Serdar Sariciftci, N. (2004). Plastic photovoltaic devices. *Materials Today*, 7(9), 36–40.

Shao, Y., Xiao, Z., Bi, C., Yuan, Y., Huang, J. (2014). Origin and elimination of photocurrent hysteresis by fullerene passivation in $CH_3NH_3PbI_3$ planar heterojunction solar cells. *Nature Communications*, 5, 1–7.

Shockley, W. and Queisser, H.J. (1961). Detailed balance limit of efficiency of p-n junction solar cells. *Journal of Applied Physics*, 32(3), 510–519.

Shur, M. (1990). *Physics of Semiconductor Devices*. Pearson, New York.

Sirringhaus, H. (2005). Device physics of solution-processed organic field-effect transistors. *Advanced Materials*, 17(20), 2411–2425.

Sirringhaus, H. (2009). Reliability of organic field-effect transistors. *Advanced Materials*, 21, 3859–3873.

Sirringhaus, H., Wilson, R.J., Friend, R.H., Inbasekaran, M., Wu, W., Woo, E.P., Grell, M., Bradley, D.D.C. (2000). Mobility enhancement in conjugated polymer field-effect transistors through chain alignment in a liquid-crystalline phase. *Applied Physics Letters*, 77(3), 406–408.

So, F., Kido, J., Burrows, P. (2008). Organic light-emitting devices for solid-state lighting introduction: The potential for. *Mrs Bulletin*, 33(July), 663–669.

Søndergaard, R., Hösel, M., Angmo, D., Larsen-Olsen, T.T., Krebs, F.C. (2012). Roll-to-roll fabrication of polymer solar cells. *Materials Today*, 15(1/2), 36–49.

Stassen, A.F., De Boer, R.W.I., Losad, N.N., Morpurgo, A.F. (2004). Influence of the gate dielectric on the mobility of rubrene single-crystal field-effect transistors. *Applied Physics Letters*, 85(17), 3899–3901.

Steiger, J., Karg, S., Schmechel, R., Von Seggern, H. (2001). Aging induced traps in organic semiconductors. *Synthetic Metals*, 122(1), 49–52.

Steim, R., Ameri, T., Schilinsky, P., Waldauf, C., Dennler, G., Scharber, M., Brabec, C.J. (2011). Organic photovoltaics for low light applications. *Solar Energy Materials and Solar Cells*, 95(12), 3256–3261.

Stranks, S.D., Eperon, G.E., Grancini, G., Menelaou, C., Alcocer, M.J.P., Leijtens, T., Herz, L.M., Petrozza, A., Snaith, H.J. (2013). Electron-hole diffusion lengths exceeding 1 micrometer in an organometal trihalide perovskite absorber. *Science*, 342(October), 341–345.

Sum, T.C. and Mathews, N. (2014). Advancements in perovskite solar cells: Photophysics behind the photovoltaics. *Energy and Environmental Science*, 7(8), 2518–2534.

Sun, Y. and Forrest, S.R. (2008). Enhanced light out-coupling of organic light-emitting devices using embedded low-index grids. *Nature Photonics*, 2(8), 483–487.

Swanson, R.M. (2006). A vision for crystalline silicon photovoltaics. *Progress in Photovoltaics: Research and Applications*, 14, 443–453.

Tanase, C., Meijer, E.J., Blom, P.W.M., de Leeuw, D.M. (2003). Unification of the hole transport in polymeric field-effect transistors and light-emitting diodes. *Physical Review Letters*, 91(21), 7–8.

Tang, C.W. (1986). Two-layer organic photovoltaic cell. *Applied Physics Letters*, 48(2), 183–185.

Taylor, D.M., Drysdale, J.A., Torres, I., Fernández, O. (2006). Electron trapping and inversion layer formation in photoexcited metal-insulator-poly(3-hexylthiophene) capacitors. *Applied Physics Letters*, 89(18), 87–90.

Todescato, F., Capelli, R., Dinelli, F., Murgia, M., Camaioni, N., Yang, M., Bozio, R., Muccini, M. (2008). Correlation between dielectric/organic interface properties and key electrical parameters in PPV-based OFETs. *Journal of Physical Chemistry B*, 112(33), 10130–10136.

Tokito, S., Tanaka, H., Okada, A., Taga, Y. (1996). High-temperature operation of an electroluminescent device fabricated using a novel triphenylamine derivative. *Applied Physics Letters*, 69(7), 878–880.

Tsutsui, T. and Takada, N. (2013). Progress in emission efficiency of organic light-emitting diodes: Basic understanding and its technical application. *Japanese Journal of Applied Physics*, 52.

Tsutsui, T., Yahiro, M., Yokogawa, H., Kawano, K., Yokoyama, M. (2001). Doubling coupling-out efficiency in organic light-emitting devices using a thin silica aerogel layer. *Advanced Materials*, 13(15), 1149–1152.

Turak, A. (2013). Interfacial degradation in organic optoelectronics. *RSC Advances*, 3(18), 6188–6225.

Tyan, Y.-S. (2011). Organic light-emitting-diode lighting overview. *Journal of Photonics for Energy*, 1, 011009.

Uchida, S., Xue, J., Rand, B.P., Forrest, S.R. (2004). Organic small molecule solar cells with a homogeneously mixed copper phthalocyanine: C 60 active layer. *Applied Physics Letters*, 84(21), 4218–4220.

Uoyama, H., Goushi, K., Shizu, K., Nomura, H., Adachi, C. (2012). Highly efficient organic light-emitting diodes from delayed fluorescence. *Nature*, 492(7428), 234–238.

Uzu, H., Ichikawa, M., Hino, M., Nakano, K., Meguro, T., Hernández, J.L., Kim, H.S., Park, N.G., Yamamoto, K. (2015). High efficiency solar cells combining a perovskite and a silicon heterojunction solar cells via an optical splitting system. *Applied Physics Letters*, 106, 013506.

Veres, J., Ogier, S., Lloyd, G., de Leeuw, D. (2004). Gate insulators in organic field-effect transistors. *Chemistry of Materials*, 16(23), 4543–4555.

Verploegen, E., Miller, C.E., Schmidt, K., Bao, Z., Toney, M.F. (2012). Manipulating the morphology of P3HT-PCBM bulk heterojunction blends with solvent vapor annealing. *Chemistry of Materials*, 24(20), 3923–3931.

Wagenpfahl, A., Rauh, D., Binder, M., Deibel, C., Dyakonov, V. (2010). S-shaped current-voltage characteristics of organic solar devices. *Physical Review B – Condensed Matter and Materials Physics*, 82(11), 1–8.

Wang, S.D., Yan, Y., Tsukagoshi, K. (2010). Transition-voltage method for estimating contact resistance in organic thin-film transistors. *IEEE Electron Device Letters*, 31(5), 509–511.

Wu, S.F., Li, S.H., Wang, Y.K., Huang, C.C., Sun, Q., Liang, J.J., Liao, L.S., Fung, M.K. (2017). White organic LED with a luminous efficacy exceeding 100 lm W^{-1} without light out-coupling enhancement techniques. *Advanced Functional Materials*, 27(31), 1–9.

Yagi, I., Tsukagoshi, K., Aoyagi, Y. (2004). Direct observation of contact and channel resistance in pentacene four-terminal thin-film transistor patterned by laser ablation method. *Applied Physics Letters*, 84(5), 813–815.

Yamasaki, T., Sumioka, K., Tsutsui, T. (2000). Organic light-emitting device with an ordered monolayer of silica microspheres as a scattering medium. *Applied Physics Letters*, 76(10), 1243–1245.

Yoo, S., Domercq, B., Kippelen, B. (2004). Efficient thin-film organic solar cells based on pentacene/C 60 heterojunctions. *Applied Physics Letters*, 85(22), 5427–5429.

Yu, G. and Heeger, A.J. (1995). Charge separation and photovoltaic conversion in polymer composites with internal donor/acceptor heterojunctions. *Journal of Applied Physics*, 78(7), 4510–4515.

Yu, L., Liu, J., Hu, S., He, R., Yang, W., Wu, H., Peng, J., Xia, R., Bradley, D.D.C. (2013). Red, green, and blue light-emitting polyfluorenes containing a dibenzothiophene-S,S-dioxide unit and efficient high-color-rendering-index white-light-emitting diodes made therefrom. *Advanced Functional Materials*, 23(35), 4366–4376.

Zhou, X., Pfeiffer, M., Huang, J.S., Blochwitz-Nimoth, J., Qin, D.S., Werner, A., Drechsel, J., Maennig, B., Leo, K. (2002). Low-voltage inverted transparent vacuum deposited organic light-emitting diodes using electrical doping. *Applied Physics Letters*, 81, 922–924.

Zhu, Z., Waller, D., Gaudiana, R., Morana, M., Mühlbacher, D., Scharber, M., Brabec, C. (2007). Panchromatic conjugated polymers containing alternating donor/acceptor units for photovoltaic applications. *Macromolecules*, 40(6), 1981–1986.

Zilker, S.J., Detcheverry, C., Cantatore, E., de Leeuw, D.M. (2001). Bias stress in organic thin-film transistors and logic gates. *Applied Physics Letters*, 79(8), 1124–1126.

Zschieschang, U., Weitz, R.T., Kern, K., Klauk, H. (2009). Bias stress effect in low-voltage organic thin-film transistors. *Applied Physics A: Materials Science and Processing*, 95(1), 139–145.

Zschieschang, U., Yamamoto, T., Takimiya, K., Kuwabara, H., Ikeda, M., Sekitani, T., Someya, T., Klauk, H. (2011). Organic electronics on banknotes. *Advanced Materials*, 23(5), 654–658.

Index

Other titles from

iSTE

in

Electronics Engineering

2021

NGUYEN Thien-Phap
Organic Electronics 1: Materials and Physical Processes

2020

CAPPY Alain
Neuro-inspired Information Processing

FOUQUET François
Noise in Radio-Frequency Electronics and its Measurement

MAHDOUM Ali
CAD of Circuits and Integrated Systems

2018

MURET Pierre
Fundamentals of Electronics 2: Continuous-time Signals and Systems
Fundamentals of Electronics 3: Discrete-time Signals and Systems, and Quantized Level Systems

2017

MURET Pierre
Fundamentals of Electronics 1: Electronic Components and Elementary Functions

BUCCI Davide
Analog Electronics for Measuring Systems

2016

BAUDRAND Henri, TITAOUINE Mohammed, RAVEU Nathalie
The Wave Concept in Electromagnetism and Circuits: Theory and Applications

FANET Hervé
Ultra Low Power Electronics and Adiabatic Solutions

NDJOUNTCHE Tertulien
Digital Electronics 1: Combinational Logic Circuits
Digital Electronics 2: Sequential and Arithmetic Logic Circuits
Digital Electronics 3: Finite-state Machines

2015

DURAFFOURG Laurent, ARCAMONE Julien
Nanoelectromechanical Systems

2014

APPRIOU Alain
Uncertainty Theories and Multisensor Data Fusion

CONSONNI Vincent, FEUILLET Guy
Wide Band Gap Semiconductor Nanowires 1: Low-Dimensionality Effects and Growth
Wide Band Gap Semiconductor Nanowires 2: Heterostructures and Optoelectronic Devices

GAUTIER Jean-Luc
Design of Microwave Active Devices

LACAZE Pierre Camille, LACROIX Jean-Christophe
Non-volatile Memories

TEMPLIER François
OLED Microdisplays: Technology and Applications

THOMAS Jean-Hugh, YAAKOUBI Nourdin
New Sensors and Processing Chain

2013

COSTA François, GAUTIER Cyrille, LABOURE Eric, REVOL Bertrand
Electromagnetic Compatibility in Power Electronics

KORDON Fabrice, HUGUES Jérôme, CANALS Agusti, DOHET Alain
Embedded Systems: Analysis and Modeling with SysML, UML and AADL

LE TIEC Yannick
Chemistry in Microelectronics

2012

BECHERRAWY Tamer
Electromagnetism: Maxwell Equations, Wave Propagation and Emission

LALAUZE René
Chemical Sensors and Biosensors

LE MENN Marc
Instrumentation and Metrology in Oceanography

SAGUET Pierre
Numerical Analysis in Electromagnetics: The TLM Method

2011

ALGANI Catherine, RUMELHARD Christian, BILLABERT Anne-Laure
Microwaves Photonic Links: Components and Circuits

BAUDRANT Annie
Silicon Technologies: Ion Implantation and Thermal Treatment

DEFAY Emmanuel
Integration of Ferroelectric and Piezoelectric Thin Films: Concepts and Applications for Microsystems

DEFAY Emmanuel
Ferroelectric Dielectrics Integrated on Silicon

BESNIER Philippe, DÉMOULIN Bernard
Electromagnetic Reverberation Chambers

LANDIS Stefan
Nano-lithography

2010

LANDIS Stefan
Lithography

PIETTE Bernard
VHF / UHF Filters and Multicouplers

2009

DE SALVO Barbara
Silicon Non-volatile Memories / Paths of Innovation

DECOSTER Didier, HARARI Joseph
Optoelectronic Sensors

FABRY Pierre, FOULETIER Jacques
Chemical and Biological Microsensors / Applications in Fluid Media

GAUTIER Jacques
Physics and Operation of Silicon Devices in Integrated Circuits

MOLITON André
Solid-State Physics for Electronics

PERRET Robert
Power Electronics Semiconductor Devices

SAGUET Pierre
Passive RF Integrated Circuits

2008

CHARRUAU Stéphane
Electromagnetism and Interconnections

2007

RIPKA Pavel, TIPEK Alois
Modern Sensors Handbook